agnès b. stories

SEIGENSHA

agnès b. stories

アニエスベーが1983年にはじめて日本に上陸してから40年を迎えた2023年。
ブランド創設者でデザイナーのアニエス・トゥルブレにとって、パリにつづく大切な場所である日本でのアニバーサリーイヤーを記念して、アニエスベーではさまざまなコンテンツを発信しました。

本書は、アニエスベーのスピリットに共鳴する俳優やミュージシャン、アーティスト達に加え、アニエスベーを愛するスタッフが登場し、彼らのアニエスベーにまつわる物語を紡いだデジタルコンテンツ「my agnès b. story」を再編集し、ブランドの歴史やこれまでの歩みを加えてまとめたものです。

"自分らしさ"を引き出すアニエスベーの服を纏って撮り下ろされた写真や、40名のキャストによるメッセージを通して、ブランドのアイデンティティを紐解きます。

アニエスベー

In 2023, agnès b. celebrated its 40th anniversary in Japan.

Japan has always been a special place for the brand's founder and designer Agnès Troublé, and to commemorate this important anniversary, agnès b. has released a variety of contents.

This book is a re-edited version of the digital content published as 'my agnès b. story'. The website featured 40 stories from actors, musicians and creatives whose mindsets resonate with the brand, as well as staff members who love agnès b.

This book also explores the brand's history to date. Through photographs of the cast wearing agnès b. clothes and their personal messages, we hope to reveal the uniqueness of all featured cast and staff, and the identity that has been driving the brand for so many years.

agnès b.

agnès b. stories

Interview & Essay

Contents

History   Contents

My story with agnès b.
HIROSHI FUJIWARA (Music Producer)

Illustration: Yuko Saeki, Text: Megumi Koyama

僕とアニエスベーの物語
藤原ヒロシ（音楽プロデューサー）

僕がアニエスベーを着るようになったのは、1986年頃。当時はスケートボード用として着ることが多かった。というのも、海外の雑誌でスケーターたちがボーダーのTシャツを着ているのを見て憧れていたから。ボーダーTシャツを求めて、アニエスベーのお店に行ったのを覚えている。他にもロゴの入ったベレー帽やショルダーバッグなんかも愛用していた。アイテムだけではなくて、ブランドの姿勢にも憧れていた。当時から変わらず、音楽カルチャーをうまく取り入れているという印象があった。そういうのってイギリスがやりそうな感じなんだけど、フランスのブランドがやっているというのも面白かった。フランスってレゲエやヒップホップなど海外の音楽を取り入れるのがはやかったし、国民性としてカルチャーがきちんと根付いてるんだと思う。思い返すと、ニューヨークに住んでいた頃、ルームメイトの友人がアニエスベーで働いていた。エドヴィージュ・ベルモアという子で当時のパンクの女王だった。アーティストがスタッフをしているのも、アニエスベーらしい。

当時、雑誌『CUTiE』で持っていた連載「HFA（Hiroshi Fujiwara Adjustment）」でも、アニエスベーのアーティストTシャツやスタイリングを紹介した。カレッジスウェットにバイカーショーツを合わせ、アニエスベーのロゴキャップを被ったり。スケートブランドではなくて、あえてフランスのブランドのアニエスベーを被るっていうのが他にはない感じで気に入っていた。あの頃は、全身をブランドで統一するスタイルも多かったけれど、アニエスベーはシンプルだからこそ他のアイテムと合わせても馴染んだ。日本で流行しはじめたのは、90年代前半のDCブランドブームが落ち着きをみせた頃だったと思う。考えられないかもしれないけど、ルイ・ヴィトンやセリーヌなど、ラグジュアリーブランドはまだ洋服をつくっていなかった。海外の洋服ってオートクチュールみたいなものだったりジャンポール・ゴルチエやヴィヴィアン・ウエストウッドみたいな過激なものしかなくて。シンプルでカジュアルな海外のブランドというのはなかった。だから、アニエスベーのシンプルなカーディガンやカットソーを上品に着こなすスタイルは、まさに先駆的だったと思う。

シンプルがゆえに、それぞれのスタイルに溶け込み個性に寄り添ってくれるブランド。当時も今も変わらずにいるのはすごいこと。普遍的なものは、変わらないんだなと改めて思う。アニエスベーのボーダーTシャツもロゴキャップもカーディガンも、今着てもきっと馴染む。40年変わらなかったように、この先も変わらないんじゃないかな。

I started wearing agnès b. around 1986. At that time, I often wore them for skateboarding because I admired the skaters in overseas magazines wearing striped T-shirts. I remember going to agnès b. stores looking for striped T-shirts. I also loved their berets and shoulder bags with logos on them. It wasn't just the items; I admired the brand's philosophy. Even back then, I had the impression that agnès b. ingeniously incorporated music culture. While it seemed like something a British brand would do, it was interesting that a French brand was doing so. France quickly embraced international music like reggae and hip-hop, and I think culture is deeply rooted in their national character. Come to think of it, my roommate's friend worked at agnès b. when I was living in New York. Her name was Edwige Belmore, and she was known as the punk queen of the time. It was very agnès b. to have artists working as staff.

In my column 'HFA (Hiroshi Fujiwara Adjustment)' in the magazine *CUTiE*, I also introduced agnès b. Artist T-shirts and logo caps. I would pair college sweatshirts with biker shorts and top it off with an agnès b. logo cap. I liked the uniqueness of wearing agnès b., a French brand, instead of a typical skate brand. Back then, wearing a full outfit from a single brand was common, but agnès b.'s simplicity made it easy to mix and match with other items. I think agnès b. started to become popular in Japan in the early 90s, when the DC boom was settling down. It might be hard to believe, but at that time, luxury brands like Louis Vuitton and Céline were not making clothes yet. The only foreign fashion available in Japan was haute couture or extreme styles like Jean-Paul Gaultier and Vivienne Westwood. There were no simple, casual foreign brands. So, wearing simple cardigans and cut-and-sew pieces from agnès b. in an elegant way was pioneering.

It is a brand that is simple, but blends into different styles and embraces individuality. It is remarkable to uphold the same belief for so long and it made me realise that timeless items never change. The striped T-shirts, logo caps, and cardigans from agnès b. blend in seamlessly with today's style. Like how they haven't changed in the past 40 years, I believe agnès b. will remain timeless and true to its essence in the future.

藤原ヒロシ　音楽プロデューサー　80年代よりクラブDJをはじめ、日本のヒップホップ黎明期に大きな影響を与える。90年代からは音楽プロデュース、作曲など活動の幅を広げる。fragment designを主宰し、ファッションやライフスタイルなどジャンルを超えた他ブランドとのコラボレーションを行うなど、日本のストリートカルチャーを牽引するひとり。

Since the 80s, Hiroshi Fujiwara has been a pioneering club DJ, significantly influencing the early days of Japanese hip-hop. In the 90s, Hiroshi expanded his activities to include music production and composition. As the founder of fragment design, Hiroshi has led Japanese street culture through collaborations with various brands across genres such as fashion and lifestyle.

アニエスベー と聞くと、どんなアイテムを思い浮かべるだろう？ 1966年にウィリアム・クラインン監督の映画『ポリー・マグー お前は誰だ』のためにボーダー Tシャツを制作した後、1977年に誰もが楽しめるアイテムとして、ラガーシャツのコットン生地 を用いたボーダー Tシャツが誕生した。日本上陸当時、世間ではシンプルとは逆行したスタイルが流行していた。そこに突如現 れたのが、アニエスベーだった。ボーダー Tシャツをはじめとしたフレンチカジュアルな服は、男女を問わずみんながとりこに。 当時のカルチャーとボーダー Tシャツについて、ファッション誌『GINZA』元編集長の中島敏子が振り返る。

What Striped T-shirts Taught Me

ボーダーシャツが教えてくれたこと
中島敏子

...at items come to mind when you hear the name of agnès b.? After creating a striped T-shirt for the 1966 film ...o *Are You, Polly Maggoo?* by William Klein, agnès b. introduced a striped T-shirt made from cotton rugby ...s in 1977 as an item everyone could enjoy. When the brand first arrived in Japan, the prevailing fashion trend ...anything but simple, and it was during this time that agnès b. suddenly appeared. The French casual wear, ...ding the iconic striped T-shirt, captivated both men and women alike. Toshiko Nakashima, former *GINZA* ...r-in-chief, reflects on the culture of that time and the significance of the striped T-shirt.

...b. made its debut in Japan in 1983. The fashion scene ...ominated by a wave of famous Italian brands entering ...rket, leading to a craze for Italian casual fashion. ... this time, the buzzwords 'Marukane/Marubi' captured ...volous and superficial nature of the era as people ...d into the bubble economy. Men were captivated by ...suous and glamorous allure of Italian fashion, while ..., wearing bodycon dresses in the spotlight, raised ...ices in excitement. While George Orwell depicted a ...an future in his science fiction novel *1984*, the reality ... Japan was quite the opposite—it was a time of frenzy ...dness, with capitalism running rampant in the opposite ...

...es b. came to Japan at such a time, and a year later, ... the first store opened. The elegant, understated, ...ality clothing did not initially make a loud statement.

However, it gradually began to attract the attention of a particular group of sophisticated individuals. Both then and now, those who genuinely love film, art, music, and books tend to be anti-conservative and anti-extravagance. For example, though it might be a biased view, these individuals would prefer films like *Koyaanisqatsi* (music by Philip Glass, produced by Francis Ford Coppola) or *Nostalghia* (directed by Andrei Tarkovsky) over *Ghostbusters* or *Jaws 3* (all released in Japan in 1984). agnès b. likely spread gradually through the cultural context and networks of these groups.

Another significant trend was born from the magazine *Olive*. The 1984 'Lycéenne Special' was a sensation. The word 'lycéenne (high school girl)' itself thrilled young girls. The fashion was effortlessly chic and clean—navy blazers, berets, cardigans, and, of course, striped shirts. At that time, agnès b. had a considerable influence.

アニエスベーが日本に上陸した1983年。世の中のファッションは、イタリアの有名ブランドが多数上陸したこともあり、イタリアンカジュアルが席巻していたように思う。「丸金・丸ビ」という軽佻浮薄な言葉に浮かれて、人々はバブルへと突入した。世の中の男性たちはイタリアファッションのたっぷりとした官能や艶やかさに夢中になり、女性たちはお立ち台にボディコン姿で嬌声をあげていた。ジョージ・オーウェルはSF小説『1984年』でディストピアな未来を描いたが、現実の1984年の日本は真逆の資本主義が暴走する、狂騒と狂乱の日々だった。

そんな時代にアニエスベーは日本に上陸して、1年後の1984年には1号店がオープンした。楚々とした品のいい洋服たちは、当初、自分から大きく声をあげることはなかったように思う。しかし、やがてじわじわと一部の粋人たちに注目されていったのではなかったか。今も昔も、映画やアート、音楽や本が本当に好きな人たちはどちらかというと、アンチ・コンサバ、アンチ・バブル（金満）だと思っている。たとえば勝手なイメージだが、『ゴーストバスターズ』や『ジョーズ3』よりは、『コヤニスカッティ』（音楽：フィリップ・グラス、製作：F.F.コッポラ）や『ノスタルジア』（監督：A.タルコフスキー）を好む人たち（以上すべて1984年に日本公開）。アニエスベーは、そんな後者のカルチャーの文脈と人脈を通してじわじわと広がっていったブランドではないだろうか。

もう1つの大きな潮流は雑誌『Olive』から生まれた。1984年の「リセエンヌ特集」は衝撃だった。まず、リセエンヌという言葉に少女たちは胸をときめかせた。さりげなく小粋で清潔感のあるファッション。紺ブレ、ベレー帽、カーディガン、そしてもちろんボーダーシャツ。当時のアニエスベーの影響力は大きかった。

歴史上、多くの文化人たちもボーダーを愛してきた。ピカソ、ヘミングウェイ、デヴィッド・ホックニー、アンディ・ウォーホル、ココ・シャネルからジャンポール・ゴルチエまで、高名な芸術家たちがそれぞれのボーダーシャツを着ていた。しかし『Olive』でボーダーといえば映画『なまいきシャルロット』（1985年）のシャルロット・ゲンズブールにつきる。彼女の母親のジェーン・バーキンもまた、ボーダーのカットソーがよく似合う人だった。ジェーンとシャルロットはOlive少女たちのアイコンで

もあった。

また当時、『Olive』で飛ぶ鳥を落とす勢いの人気だったフリッパーズ・ギターも、ボーダーに愛された芸術家たちだった。2人の着るフレンチファッションは、彼らの音楽を愛する男女のファンどちらも魅了した。男の子も女の子も同じ服で何が悪い。今でいうところのジェンダーレスファッションを、ボーダーを通してこの時代にすでに軽やかに表現していたのは、雑誌と音楽だった。こうしてユースカルチャーを通じて、ボーダーは世の中に静かに浸透していった。

何回も組まれた『Olive』のリセエンヌ特集のある号に、モデルではなく普通の女の子たちが全員ボーダーシャツを着て楽しげに並んでジャンプしている写真がある。それはボーダーが日本の女の子たちのスタンダードなアイテムになったことを意味していた。シンプルなだけに、誰が着るかでまったく表情が変わる服。それは女の子たちに、ファッションも生き方もどちらも「自分で考える自主性と個性が大事だよ」と伝え続けた『Olive』の精神とぴったり重なるものだ。学歴や肩書きや性別よりも、もっと大事なことがあるよ。懐かしい1枚のボーダーシャツに袖を通すたびに、昔の気持ちをふと思い出す。

ところで私は、2011年から7年間、雑誌『GINZA』の編集長を務めていた。その間に『GINZA』の企画として『おとなのオリーブ』という別冊を作ったことがある。若い読者にも話題になったが、特に元Olive少女たちが見つけてくれてSNSを通して口コミで話題になり、最後は大変な反響になった。あちこちでトークショーをしたり、即席の「オリーブカフェ」を作ったりしたが、どの場所にも必ず何人かボーダーを着た妙齢の女性がいて、友人とペアルックだったりする姿が微笑ましかった。

当時、毎年パリコレのショーで見るアニエスベーの洋服は常にエターナルな少女の夢に溢れていて、私はいつも密かに『Olive』を思い出していた。でも実はショーの最後に出てくるアニエスさんが一番可愛らしく、永遠の少女そのものだったのだ。

Historically, many cultural figures have loved stripes. Renowned artists such as Picasso, Hemingway, David Hockney, Andy Warhol, Coco Chanel, and Jean-Paul Gaultier wore striped shirts. However, when it comes to stripes in *Olive*, it is epitomised by Charlotte Gainsbourg in the film *L'Effrontée* (1985). Her mother, Jane Birkin, also looked great in striped tops. Both Jane and Charlotte were icons for the girls who read *Olive*.

At that time, the Flipper's Guitar, wildly popular in *Olive*, was also an artist group loved by stripes. The French fashion worn by the duo captivated both male and female fans who loved their music. What was wrong with boys and girls wearing the same clothes? Through stripes, the magazines and music of this era already lightly expressed what we now call genderless fashion. In this way, through youth culture, stripes quietly permeated society.

In one of the issues of *Olive*'s 'Lycéenne Special' feature, there is a photo of ordinary girls, not models, all wearing striped shirts, joyfully jumping in a row. This signified that stripes had become a standard item for Japanese girls. The simplicity of the shirt allows it to change its expression entirely depending on who wears it. This aligns perfectly with the spirit of *Olive*, which always conveyed to girls that both fashion and lifestyle should value individual thought and personal uniqueness. It emphasised that there are things more important than academic background, titles, or gender. It brings back those old feelings every time one slips into a nostalgic striped shirt.

By the way, I was the editor-in-chief of *GINZA* magazine for seven years from 2011. During that time, as a project for *GINZA*, I created a special issue called 'Adult Olive'. It became a topic of conversation among young readers, but especially among former *Olive* girls, who discovered it and spread the word through social media, leading to an overwhelming response. We held talk shows in various places and created impromptu 'Olive cafés'. At each location, there were always a few stylish women wearing stripes, sometimes even matching outfits with their friends, which was heartwarming.

At that time, the clothes I saw at the agnès b. shows during Paris Fashion Week were always filled with the eternal dreams of young girls, and they always secretly reminded me of *Olive*. But it was Agnès herself, who would appear at the end of the show, who was always the most charming, embodying the essence of an eternal young girl.

中島敏子　編集者、プロデューサー　リクルートを経てマガジンハウスに入社。『BRUTUS』『Tarzan』『relax』にて副編集長。カスタム出版部を経て、2011年から2017年まで『GINZA』編集長に。2014年度「世界のファッション業界を形成する500人」に選出。現在はフリーランスとして編集、プロデュースをつとめる。

After starting her career at Recruit, Toshiko Nakashima joined Magazine House. Toshiko served as deputy editor for *Brutus, Tarzan*, and *relax*, and later led the Custom Publishing Department. From 2011 to 2017, Toshiko was the editor-in-chief of *GINZA*. In 2014, she was named one of the '500 People Shaping the Global Fashion Industry'. Currently, Toshiko works as a freelance editor and producer in the fashion and culture fields.

HIKARI MITSUSHIMA

架け橋となるものづくり
満島ひかり

Photography: Yuichiro Noda
Hair & Make-up: Naoki Ishikawa
Text: Rei Sakai

Connecting
the Future

Interview

Interview

agnès b. stories

14

いつまでも子どものようにピュアでいられたら……。デザイナーのアニエス・トゥルブレは、
いつまでも少女のような心で、今もなおクリエイションを続けている。自宅で見つけた落ち葉や、
通勤途中で出会う街並み。いつもと変わらない景色も、彼女のまなざしを通せばたちまち生地や服となる。
2023年にレーベル「Rhapsodies」を立ち上げた、俳優でアーティストの満島ひかり。
彼女もまたピュアな感覚を頼りにクリエイトをする表現者のひとり。
ものづくりの難しさや楽しさを体現してきた彼女が見つけた、あたらしい居場所とは。

If I could be as pure as a child forever... Designer Agnès Troublé continues to create with a heart as innocent as a young girl's. Fallen leaves found at home, or the streetscapes she encountered during her commute... Through her eyes, even the most ordinary views instantly transform into fabrics and clothing. Actor and artist Hikari Mitsushima, who launched the label 'Rhapsodies' in 2023, is also a creator who relies on her pure sensibilities in her creations. What is the new place she has found, that embodies the challenges and joys of creating?

「ものに溢れた世界の中で、失われていく感性もあって。良いものまで消えないように、繋げていきたい」。そう話す満島さんは、2018年にフリーランスになり、ひとつの作品をつくりあげるまでの道のりの長さを改めて体感していたところだった。「新しく何かをつくるのなら、これまでとこれからを繋ぐエネルギーでありたい。固定概念を外して清らかでいたいんです」という彼女。

ものづくりの過程を自分の目で確かめて、一つひとつを美しいものにしていきたい。そんな想いに突き動かされてクリエイションレーベルを立ち上げた。「みんながどんなことを表現するのか、肩書きとか関係なく気になる人がたくさんいます。あの人のああいうところが素敵、この人すごい! なんて。どんどん気持ちをリッチにしてフラットなものづくりをしたい、主人公は私でなくてもいいと思っているんです」。

数々の作品で主演をつとめてきた満島さんが、新しいものづくりのプラットフォームに選んだのは、過去から未来へと繋ぐ"橋渡しの役"だった。「表に出たり裏に回ったりして、呼吸しやすい環境や感性を守って広げていけたら良いな。みんなが自然と繋がっていられるなら、それが最も気楽で幸せだけど、余裕のない人や流れを予測できないことも沢山あるし、きっと自分にもそういうタイミングってある。大所帯になればなるほど、意識的に繋げていくことも必要だなって」。レーベル初の楽曲『eden』では、彼女が作詞を担当し、全体の総指揮をとった。テーマとなった楽園やプリミティブという感覚も彼女自身が生まれ持ったもの。「原始的で真っさらで、定まらない感性。以前は直感も大切にしていたけど今は、直感より手前で決めることにしています。大人になると直感すらコントロールできる気がして」。

何でも焼き増しのように量産されてしまう世界のイビツさに反発していくように、楽曲内には広々とした感性が言葉や音となって散りばめられている。「暇や無駄にも見える、ふわふわした時間を愛おしく感じます。エンターテインメントと呼ばれるものには疑わしい部分もありますが、良いチカラもあるとまだ信じているので、生活にユーモアを持って小さな感動を重ねて、遊びや才能をのびやかに活かしたいです。歳を重ねて知を得ることの素晴らしさや、幼さゆえの多感さを持つ色んな人々と、豊かに共存して制作していきたいです」。

'In a world overflowing with things, there is a certain sensibility that seems to be disappearing. I want to keep the good things from vanishing and for them to connect to the future,' says Hikari, who became independent in 2018—she had just experienced the long journey of creating a single work again. Hikari continues, 'If I'm going to create something new, I want it to be an energy that connects the past and the future. I want to create with purity, free from preconceived notions.'

Hikari wants to personally ensure the creation process and make each piece beautiful. Driven by these feelings, she started a creative label. 'There are so many people with intriguing creations, regardless of their titles. Like, "This person has great qualities in this specific area, or this person is amazing!" I wanted to continuously enrich my own feelings and create on an equal footing, and I didn't have to be the protagonist.' From the beginning, she was interested in the people who support creation behind the scenes.

Hikari, who has starred in numerous works, chose to be a 'bridge' connecting the past to the future in this new creative platform. 'I want to create an environment where everyone can breathe easily, where sensibilities are preserved and expanded, whether I'm in the spotlight or behind the scenes. While it's best if everyone can connect naturally, people might not have the time or leeway, or there can be unpredictable situations. There are times when I am like that too. The larger the group of people, the more important it is to stay consciously connected.'

Hikari wrote the lyrics for the label's first song, 'eden', and oversaw the entire project. The themes of paradise and the primitive senses are elements she was born with. 'Primitive, fresh, and undefined sensibilities. I used to value intuition, but now, I try to decide even before intuition jumps in. I feel that we can even control our intuition as adults.'

In defiance of a world where everything is mass-produced like photocopies, an expansive sensibility expressed through words and sounds is scattered in her songs. 'I think the ambiguous moments that are seemingly idle or wasteful are very important. While there are questionable aspects to what we call entertainment, I still believe in its positive power. I want to bring humour to life and accumulate these small moments of wonder, freely harnessing playfulness and talent. As we grow older, we see colours more clearly, and we gain knowledge and understand people's feelings a bit too well. I want to create and co-exist with people who have gained rich experiences and wisdom as we get older, and also with people who have the sensitivities of youth.'

満島ひかり　俳優,アーティスト　俳優を中心に、音楽や執筆など多彩に活躍。2023年にクリエイションレーベル「Rhapsodies」を立ち上げ、楽曲『eden』『shadow dance』を発表。書籍『軽いノリノリのイルカ』(マガジンハウス)、ラジオ『ヴォイスミツシマ』。アニメ「アイラブみー」では60役以上の声を担当する。

With a focus on acting, Hikari Mitsushima is also active in various fields, including music, and writing. In 2023, Hikari launched the creative label 'Rhapsodies' and released the songs 'eden' and 'shadow dance'. She published *Karuinorinorinoiruka* (Magazine House), and hosts the radio show 'Voice Mitsushima'. She voices over 60 characters in the animation 'I Love Me'.

Interview

これまで数々のアイコンを生み出してきたアニエスベーのバッグコレクション。そのデザインの一端を担う彼がはじめてブランドと出会ったのは中学生の頃。「当時から憧れていた」という創設者でありデザイナーのアニエス・トゥルブレとの出会いと、共に仕事を進める中で得たものとは。

　「ファッションが好きになった中学生時代。アニエスベーと言えばコーネリアスの小山田圭吾さんや藤井フミヤさんはじめカルチャーアイコンとなっている人たちがこぞって着ていたこともあって、とにかく影響を受けましたね。友人が白黒のボーダーTシャツを着ていて、それがすごく似合っていて羨ましかった。当時は、人と被るのが嫌で敬遠していたけれど、大人になってようやく自分なりに着こなせるようになった気がします。それこそ、大学の入学式はアニエスベーのセットアップを着て行ったんですよ。懐かしいですね」

　そう話すのは、これまで10年以上に渡りアニエスベーのバッグを手掛けてきたデザイナー。京都造形芸術大学（現・京都芸術大学）で服づくりを学び、現在はバッグをはじめ多様なアイテムのデザインを行っている。バッグのデザインは、フランス・パリにいるアニエス・トゥルブレ本人との共同作業。毎シーズン彼女の元を訪れ、さまざまな議論を重ねた上で一つひとつのバッグを作り上げていくのだという。

　「コロナ禍もあってなかなか直接会えなかったけれど、昨年久々にパリでミーティングをして。ファッションデザイナーとして世界中で知られた存在ですが、すごくフランクでピュアな方。彼女はいつも世の中のトレンドではなく、自分自身の体験に基づいたインスピレーションを大切にしている。道で出会った景色や出来事、映画や音楽。すごく正直だなと感じます。デザインも僕の想像にはない絵を描くので、毎回「そう

くるか……!」という驚きがあります。話していたかと思うと急にそばにあった裏紙にスケッチを描き始めたり、ある時はたくさんの落ち葉をテーブルに広げて色の話をしたり。僕たちは毎回あたふたさせられるけど（笑）。もう10年以上バッグのデザインに携わっていますが、会う度に新しいインスピレーションがあって、デザインプロセスも常に変化していきます」

　ブランド設立当初からシンプルでタイムレスなデザインを生み出し続けるアニエス・トゥルブレ。フランス特有のエレガンスを纏いながら、実用的かつエモーショナルなフィーリングを感じさせるコレクションは、多くのインスピレーションと細部に渡るこだわりによって支えられている。

　「一緒に働くようになって、それまで漠然と憧れていたものにここまでのこだわりがあるのかということに驚きました。アニエスベーのコレクションはシンプルなアイテムが多いのですが、一見すると普遍的なアイテムも、着るとスタイルを与えてくれる。自分に必要なものやクオリティを理解している人が着ると、スッとハマるというか。ものづくりへの細かなこだわりがあるからこそ、ベーシックな装いもアニエスらしい佇まいになる。憧れを作る一方で、誰にでも寄り添ってくれるような懐の深さは、そうしたところから生まれているんだなと思います」

　1975年にパリで創業してから今日まで、日々自身が得たインスピレーションをファッションという形で表現する。ルーティンではないその

Inspiration is Always Nearby

ヒントはすぐそばに

フォースアイ

'I started to like fashion when I was in junior high school. agnès b. greatly influenced me, especially because cultural icons like Keigo Oyamada from Cornelius and Fumiya Fujii were all wearing them. That really impacted me. At that time, I avoided wearing the same things as others, but as an adult, I finally felt like I could style agnès b. in my way. In fact, I wore an agnès b. suit for my university entrance ceremony. Those are fond memories.'

These words come from a designer who has been creating bags for agnès b. for over 10 years. After studying fashion at Kyoto University of Art and Design, FORCE EYE now designs various items, including bags. The bag designs are a collaborative effort with Agnès Troublé herself in Paris. Each season, FORCE EYE visits Agnès, and through numerous discussions, they meticulously craft each bag together.

'Although a globally renowned fashion designer, Agnès is incredibly down-to-earth and genuine. Agnès always values inspiration from her personal experiences over societal trends. It could be a scene she encountered on the street, an event, a movie, or music. I find her honesty refreshing. Her designs often surprise me because they are

beyond my imagination, making me think, "So that's how she sees it!" She might start sketching on a scrap of paper while we're talking, or spread out a bunch of leaves on the table to discuss colours. We are always kept on our toes! (Laughs) Every meeting with her brings new inspiration, and our design process continually evolves.'

agnès b.'s collections embody a uniquely French elegance while evoking practical and emotional feelings, supported by a wealth of inspiration and meticulous attention to detail.

'Working with her, I was amazed to discover the level of detail and dedication behind what I had previously admired from afar. While agnès b. collections feature many simple items, these seemingly universal pieces transform into something stylish when worn. They fit seamlessly with those who understand their needs and quality. The meticulous attention to craftsmanship makes even basic attire embody the essence of agnès b. While creating objects of admiration, the brand also possesses a profound inclusiveness that resonates with everyone. I believe this depth comes from such dedication.'

The agnès b. bag collection has produced numerous iconic designs over the years. FORCE EYE, the designer behind some of these creations, first encountered the brand in junior high school. Here, we explore his meeting with founder and designer Agnès Troublé, whom he has admired since then, and what he gained from working alongside her.

姿勢を継続しているアニエスに教えられることはとても多い。

　「ものすごく元気で、あのエネルギーはどこからくるんだろうって。僕の倍以上生きている大先輩が、今なお一つひとつにこだわりを持って自分自身でデザインを手掛けている。その姿勢やエネルギー、精神力は本当にリスペクトしています。人間なんで飽きたり、疲れたり、ルーティンになってしまうことってあると思うけれど、彼女はそれを全く感じさせない。いつだって美しさやかっこよさ、可愛さというものに一喜一憂できるというのはすごいことです。アニエスさんと会うとたくさんの刺激をもらえる一方で、襟を正されるような気持ちになる。それに、自分で作っている服だから当たり前だと思われるかもしれませんが、アニエスベーの羊服は、彼女が1番おしゃれに着こなしていると思っていて。それは、感覚が若くていつもファッションを彼女なりの哲学をもって楽しんでいるからだと思うんです。90年代など、新しくてアヴァンギャルドなものがもてはやされた時代にも、一貫してシンプルな服を作り続けてきた。それを貫く姿勢こそロックですよね。中学1年生の頃からファッションデザイナーとして憧れていた人が、今も変わらない哲学をもっていて、自分がそこに携われている。こんなに嬉しいことはないですね」

自宅やオフィスには、ルネ・ガブリエル、オラヴィ・ハンニネン、マルセル・ガスコアン、ピエール・ジャンヌレ、ジャン・プルーヴェなどのデザイナーが手掛けた名作椅子や家具のほか、数々のアート作品が。自身の興味の幅が狭くならないように、あえてすべてに統一感を持たせないよう意識しているそう。

In his home and office, FORCE EYE has chairs and furniture designed by renowned designers such as René Gabriel, Olavi Hänninen, Marcel Gascoin, Pierre Jeanneret, and Jean Prouvé, along with numerous art pieces. FORCE EYE consciously avoids creating a uniform look to ensure his interests remain broad and diverse.

「中学生の頃に1度手にして、ボロボロになるまで着ていた」というアーティストTシャツは、アニエスベーが英国の現代アーティスト、ギルバート&ジョージをフィーチャーしたもの。「今日着ているのはその後の復刻版。懐かしくてすぐに買いました。大人になってまた着れるとは(笑)」

The Artist T-shirt that FORCE EYE once owned in junior high school and wore until it was completely worn out was a piece featuring the British contemporary artists Gilbert & George. 'The one I'm wearing today is the reissued version. I bought it immediately out of nostalgia. It's amusing to think I can wear it again as an adult!' (Laughs)

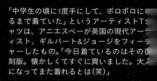

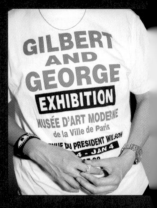

「背負っていて、良い意味で主張しない佇まいがポイント」というバックパックは、自身がデザインを手掛けたプロダクト。洋服と馴染みの良いコットンライクなナイロンを採用。随所にレザーの切り替えを施すことで落ち着いた印象に仕上げた。

The backpack FORCE EYE designed is characterised by its subtle presence. 'It makes a statement in a good way without being overly assertive.' Cotton-like nylon that pairs well with clothing is used. The leather accents give it a sophisticated look.

ブランドを象徴するボーダーTシャツは、定期的に買い足すスタンダード。「さまざまなブランドがリリースしていますが、アニエスベーのそれは柄のピッチや着丈、生地感も含めて1番安心感があります」。日本はもちろん、パリ出張の際に現地で購入することも多いそう。

The brand's iconic striped T-shirt is a staple that FORCE EYE regularly repurchases. 'Although many brands release similar items, agnès b.'s T-shirt offers reliability in terms of stripe pitch, length, and fabric feel.' FORCE EYE often buys them in Japan and during business trips to Paris.

フォースアイ　アニエスベー バッグデザイナー　10年以上に渡ってアニエスベーでバッグデザイナーとして活躍。バッグのほか、財布やポーチ、レザーグッズなども担当する。音楽、映画をはじめとした多様なカルチャーに精通。特にアートが好きで日々展覧会に通いながら見識を深めている。

For over 10 years, FORCE EYE has been a bag designer at agnès b. In addition to bags, he also designs wallets, pouches, and leather goods. He is well-versed in various creative

The Joys of Adulthood

KYOKO
KOIZUMI

大人になる楽しみ
小泉今日子

Photography: Yasutomo Ebisu
Styling: Mana Yamamoto
Hair: ASASHI
Make-up: Asami Taguchi
Text: Tomoko Ogawa

1982年に、16歳で歌手デビューして以降、俳優、執筆家、そして、制作会社「明後日」の代表取締役として
プロデュース業も務める小泉今日子。より自由に、自分らしく生きるために走り続けてきた彼女が、
80年代と原宿のカルチャーやファッション、アニエスベーとの関係を語る。
わからないなりに冒険を続けた日々がつないでくれた、もっと楽しい今のこと。

アニエスベーが日本に上陸したとき、どのような印象を持っていましたか？

K　かわいらしさと清潔さがありながらも、シンプルですごくおしゃれという感覚でした。アニエスベーが上陸する以前の10代って、カリフォルニア的なファッションに影響を受けていた世代だと思うんです。ちょうどその頃、マガジンハウスの雑誌『Olive』が「リセエンヌ」という言葉を言い始めて。アメリカ派から、パリ派、ロンドン派みたいに広がっていっていたときに出合ったのが最初ですね。一番アニエスベーを着ていたのは、80年代の終わりから90年代頭だったと思います。20歳前後の頃は、パリにお家を持っている先輩方がいたので、お休みが長く取れるとパリを訪ねて、レアール地区の1号店にも行きました。日本に入っていないアイテムを見つけると、「あ、これほしい！」と思って買っていたものです。それだけじゃなく、すごく早い段階から、さまざまなジャンルのアーティストや映画とコラボレーションをされていたので、アニエスベーを追っていると、「この映画見なくちゃ！」とか「この個展は見に行かなきゃ！」とか、そういうカルチャーをいつも教えてもらっていました。

当時は、どんなふうにコーディネートを楽しんでいたのでしょうか？

K　遊ぼうと思えば自由に遊べる服だったので、ボーダーのTシャツを着て、真っ赤な口紅をつけたり、パールのネックレスをして、ホワイトジーンズを合わせたりしていました。あとは、黒地に小花柄のプリントが入ったミニのワンピースが気に入っていて、上に白い革ジャンを着たり、足元はエンジニアブーツとか、ぶかぶかのスニーカーを合わせたり。ソフトなアイテムとハードなものを融合させるスタイルが、自分の中でちょっとクールなイケてる女という印象があって、好きだったんですよね。

女性だからといって、かわいいだけじゃない。かっこいいとも共存するスタイルは、確かにクールですよね。

K　「女の子ではいたいけれど、女の子扱いはされたくない」みたいな気分に、アニエスベーはピッタリだったんですよね。それ以前は、「男性のためにかわいくするわけじゃない」という提案が、あまりされてはいなかったんだと思います。例えば、80年代のアメリカンファッションの場合、女性のアイテムというと、肌の露出が多いとか、ややセクシーな方向に振れてしまう印象があって、多くの人が着やすいものではなかったというか。でも、パリから上陸したアニエスベーは、少しだけ不良娘っぽさやマニッシュさがあって、同時に女性っぽくもある。そのちょうどいいバランスが、日本人の生活にすごくマッチしたのだと思います。特に80年代は、男女雇用機会均等法が成立して、状況はすぐには変わりませんでしたが、日本でも女性が働くことが尊重され始めた時代でもあったので、当時の働く女性たちがアニエスベーのファッションを愛していたように私には見えていました。

80年代の小泉さんは、ファッションで何を表現していたと思いますか?

K　世の中に対して、何かケンカを売っていたみたいな（笑）、そういう気分に対して、強い味方になってくれるのが、やっぱりファッションでしたね。その頃の私には、自分の気持ちを表現するときに武装みたいなかたちで助けてくれるものとして、ファッションがあったかな。まあ、攻めていたかったんでしょうね。だから、すごく攻めてました。ありがたいことに「じゃあ、私もそういうコーディネートをしてみよう」と後をついてきてくれるフォロワーさんたちも結構いて。そういう環境だったので、ファッションでは本当にたくさん冒険しましたね。いろんな雑誌の撮影で、あらゆることに挑戦していたし、後輩から「コイズミさんって、着たことない服ないでしょ?」と言われるぐらい、何でも受け入れてました。それこそ、（服を）着ないってことまでね。ある意味、プロの着せ替え人形みたいなものでしたよね。もちろん、嫌じゃなかったし、楽しめていたから、やれていたんですけどね。

当時は原宿に住んでいらした小泉さんですが、思い出の場所はありますか?

K　当時の原宿と言えば、やっぱり、ラフォーレ原宿なんですよ。何でも入っていたから。あと、セントラルアパートの地下に、原宿プラザというロンドンのケンジントン・マーケットみたいな小さなお店がひしめき合った場所もあったし。少し脇道に入ると、下町感が残っていて、いつも買っていた八百屋さんや商店があって、アパートやお家の前に鉢植えが並んでいるような変わらない風景があるのもいいんですよね。そういうところがすごく好きだった一方で、ちょっと青山方面へ行けば、コム デ ギャルソンやKENZOといった、もう少し大人のブランドも並んでいて。どのタイミングでそっちへ足を踏み入れようかと思っていたときに（笑）、まさにその中間点にアニエスベーがあって。そんな自分を受け入れてくれた感じがありました。

原宿に行くと、当時のことを思い出したりもしますか?

K　私、10年間ぐらい、雑誌『SWITCH』で「原宿百景」という連載をやっていたこともあって、同じ場所を定点観測していると、街もやっぱり生き物なんだなと思えてくるんですよね。特に、人が集まる場所は、そうやって新陳代謝をしながら生きてるんだ、とハッキリと感じたときがあって。その感覚が、すごく面白くて。なので、昔住んでいた懐かしい青春の場所というよりは、ずっと観察していたい対象になっちゃいましたね。だって、「原宿」と呼ばれてますけど、それも通称で、もうそんな町名はなくなってしまっている。もっと歴史を遡って調べていくと、ワシントンハイツという在日米軍施設があったことも知って。

もはや、社会学者のように観察されているんですね。

K　そうやって一つの場所を観察していくと、いろんな話が聞けるから、すごく面白いと思っていて。今、裏原宿の辺りは、空き物件も増えてきてしまったけれど、そういう状況もまたどこかで変化するのかなー、とか想像しながら眺めています（笑）。

80年代の原宿を中心としたカルチャーについて、今振り返ると、どのように感じますか?

K　私たちが育ってきたのは、本当に変化と進化の時間でしかなかったと思います。戦後、約20年後に生まれて、高度成長期が落ち着いてきて、豊かになる可能性しかない世界で、どんどん新しいものが発明されて、そういうものを全部初めて見て、体

験する世代として、私たちがいたんですよね。だから、作り出すということへの、躊躇することのないエネルギーがすごく高くて、どの業種もすごく勢いがありました。毎日がパーティーで、お祭りという感覚があったんじゃないかなと思うし、特に、ファッション業界は、その中心が原宿にあったんだろうなと。街の特性としても、縦だけじゃなく、横のつながりで何かが生まれていくという感覚をずっと持っている場所だと思うんですよね。私たちよりも前の世代の人たちも喫茶店に集まって、クリエイターたちが出会い、そこで何かが生まれていたし、今も変わらずそうなんじゃないかな。例えば、裏原宿では、NIGOくんやジョニオ（高橋盾）くんや藤原ヒロシさんたちが、横のつながりで集まっていたし、MILKのディレクターの大川ひとみさんも、変わらずそこにいるビッグマザーのように若い子たちをつなげてくれて、場所を与えていった。そういう面白いカルチャーがたくさん生まれていた。そこで仲良くなった人たちとは、つかず離れずの関係だけど、「何かあったら行くよ」という気持ちはみんな持って生きてるかもね。

80年代から活動を続けてきて、どんなところが変わって、どんなところは変わらないと感じますか?

K　「歳を取ったね」と言われることはあっても、「変わったね」とは言われないんですよね。だから、何も変わっていない気もする。でもなんだろう、脳内がだいぶ大人になっているので、若い頃よりは人生が面白いです。自分と付き合うことが上手になったというか。経験が増えてくると、発想と行動がスッとつながるので。若い頃は、頭で思っていても、それをどうやって実現したらいいのかわからなくてイライラしたものですが、今は困ったことがあったら、「これはこうしたら解決できるな」とか、「あの人に話をすればかたちになる」ということがわかる。あと、昔は、自分が突っ走っていくことによって、そのエネルギーの余波が人に当たってしまうんじゃないかという不安もありましたが、今はそれも制御しながらできている、みたいな感覚があって。だから、若い頃よりずっといいですね。

自分をより知ることで、世の中も見えてくるみたいなところはありますよね。

K　自信を持って、「これが好き!」って言えるというか、それを何かのかたちにしていくことができていますよね。昔は人から、「こういうのやってみたら?」と言われて、「あ、それ面白いね。私だったらこうする」みたいに行動していたけれど、今は最初に指令する人にもなれているなと。

株式会社明後日の代表取締役であり、作品のプロデュースもされているわけですもんね。念願だった小説『ピエタ』の舞台化の公演はいかがでしたか? また、これからのプロジェクトについても聞かせていただけますか?

K　『ピエタ』は、会社の設立当初からずっとやりたくて、でもいろんな事情でなかなかできなかった作品だったんです。だから、これが失敗したら、本当に舞台をプロデュースすることはもうやめようと思っていました。でも、久しぶりの成功体験になったというか、うまくできた気はしています。この舞台化が最初の目標でしたし、わからないなりに全力疾走してきたので、一旦休みたい気分なんですよね。今、ここで初めて明後日が生まれた、くらいの感覚なので、この後何がやりたいのかをゆっくり考えて、またそこからスタートしようと考えています。

Since her debut as a singer at the age of 16 in 1982, Kyoko Koizumi has worked in acting, writing, and serving as the CEO of her production company 'asatte' (meaning 'the day after tomorrow' in Japanese). In her pursuit of living more freely and authentically, Kyoko reflects on the culture and fashion of the 80s in Harajuku, as well as her relationship with agnès b. Kyoko speaks about the ways to enjoy the present even more, resulting from the adventures she continued to explore despite uncertainties.

What impression did you have when agnès b. first came to Japan?

K   It felt cute and clean, but also very simple and stylish. Before agnès b., teenagers in Japan were heavily influenced by Californian fashion. Around that time, the magazine *Olive* from Magazine House started using the term 'Lycéenne (high school girl)'. That marked a shift from the American style to Parisian and London styles, which was when I first encountered agnès b. I think I wore agnès b. the most from the late 80s to the early 90s. When I was around 20, I had friends who had homes in Paris, so whenever I had a long break, I would visit Paris and go to the first agnès b. store in Les Halles. When I found items that weren't available in Japan, I would really want to buy them. Beyond the clothes, agnès b. had been collaborating with various artists and films from a very early stage, so following agnès b. always kept me informed about which movies to watch or which exhibitions to visit.

How did you enjoy coordinating your outfits back then?

K   The clothes were versatile and allowed us to play around freely.

I would wear a striped T-shirt with bright red lipstick, a pearl necklace, and white jeans. Another favourite of mine was a black mini dress with small floral prints, which I paired with a white leather jacket and mixed with engineer boots or oversized sneakers for footwear. I loved blending soft items with more rugged ones—it created a style that felt cool and gave me the impression of being a bit of a chic, edgy woman.

Girls aren't just cute—the idea of a style that isn't just cute but also cool is certainly very stylish.

K   'I want to be a girl, but I don't want to be treated like one' was the sentiment that agnès b. perfectly captured. Before that, there wasn't a trend for women to say, 'I'm not dressing to look cute for men'. For example, in 80s American fashion, women's items often leaned towards showing a lot of skin or had a more overtly sexy vibe, which wasn't always comfortable for everyone. But agnès b., which came to Japan from Paris, had a slight rebelliousness and mannishness while still being feminine. This perfect balance resonated well with Japanese lifestyles. The 80s in Japan was also when the Equal Employment Opportunity

Law was established. Although changes weren't immediate, it marked the beginning of an era where women's work was increasingly respected. Working women of that time seemed to love agnès b.'s fashion because it aligned with their emerging sense of identity and style.

What were you expressing through fashion in the 80s?

K   I wanted to challenge the world in some way. (Laughs) And fashion was my strong ally. When expressing my feelings, fashion was like an armour that helped me. I wanted to be bold and daring, and I was very aggressive. Thankfully, I had followers who would think, 'I'll try that coordination too.' It was an environment where I could experiment with fashion. I tried everything in various magazine shoots, and younger colleagues would say, 'Koizumi-san, there's no outfit you haven't worn, right?' I accepted everything. Including not wearing anything at all. I was like a professional dress-up doll. But it wasn't something I disliked; I enjoyed it, so I did so.

You lived in Harajuku in the 80s. Are there any memorable places for you?

K   When I think about Harajuku back then, it's definitely Laforet HARAJUKU. It had everything. There was also Harajuku Plaza, which was in the basement of the Central Apartments. It was like a smaller version of Kensington Market in London, filled with little shops. If you took a side street, you'd find an old-town feel, with greengrocers and shops where I always bought things, and the unchanged scenery of potted plants in front of apartments and houses. I loved those aspects, but if you walked towards Aoyama a little bit, you would see more mature brands like Comme des Garçons and KENZO. At the time I was thinking about when I would venture into that side of the world, agnès b. was right in the middle of those two extremes. (Laughs) It felt like it welcomed that phase I had.

Do you reminisce about those days when you go to Harajuku?

K   I used to do a series called 'Harajuku Hyakkei (100 Views of Harajuku)' for the magazine SWITCH for about 10 years, where I observed the same spots consistently. It made me realise that a city is like a living organism. Especially in places where people gather, you can see this clear sense of renewal and vitality. That feeling is incredibly fascinating. So, rather than being a place of nostalgic youth for me, Harajuku has become more of a subject I constantly want to observe. Even though it's called 'Harajuku', that's just a common name we use; the official town name doesn't exist anymore. If you delve further into its history, you'll discover that there used to be a US military facility called Washington Heights.

Your observation is almost like that of a sociologist.

K   When you observe a single place consistently, you hear various stories, which I find interesting. Recently, the number of vacant properties around Ura-Harajuku has increased, but I wonder if that situation will change again at some point. I watch it with that sense of curiosity and imagination. (Laughs)

How do you feel about the culture centred around Harajuku in the 80s when you look back now?

K   We grew up in a time of constant change and progress. Born about 20 years after the war, during a time when high economic growth settled in, we lived in a world filled with potential for prosperity. New things were being invented all the time, and we were the first generation to see and experience all these things for the first time. Because of this, there was a great, unhindered energy towards creation, and every industry was thriving. It felt like every day was a party, a festival. The fashion industry was at the heart of this, especially in Harajuku. The area is unique in fostering vertical and horizontal connections, constantly giving birth to new ideas. Even before our generation, creators would gather in cafés, meet, and create something new together. I believe that hasn't changed. In Ura-Harajuku, for example, NIGO, Jun Takahashi, and Hiroshi Fujiwara would gather through their lateral connections. Hitomi Okawa, the director of MILK, has always been there like a big mother, connecting and giving space to the younger generation. This created a lot of interesting culture. We might not be close all the time, but the people we got to know during that time shared a mutual understanding. 'If something happens, I'll be there for you.'

How do you feel about what has changed and remained the same since you started your career in the 80s?

K   People might tell me, 'You've aged', but they rarely say, 'You've changed'. So, it feels like nothing has really changed. However, my mind has matured quite a bit, making life more interesting than when I was younger. I've become better at dealing with myself. As you gain more experience, your ideas and actions connect more smoothly. When I was younger, I would get frustrated because I didn't know how to realise my ideas, but now, if there's a problem, I can think, 'I can solve this by doing that', or 'I'll talk to that person to make it happen'. Also, in the past, I worried that my energy might unintentionally affect others when I was pushing forward, but now I feel I can control it better. So, life is much better than when I was younger.

It seems like you can also see the world more clearly by knowing yourself better.

K   Yes, you can confidently say, 'I like this!' and make it happen. In the past, someone would suggest, 'Why don't you try this?' I would think, 'Oh, that's interesting. If it were me, I would do it like this'. But now, I've also become the one who initiates things.

As the representative director of 'asatte', you are also involved in producing works. The long-awaited stage adaptation of the novel *Pieta* must have been quite an experience. How was it, and could you tell us about your future projects?

K   *Pieta* was a piece I've wanted to bring to life since I established the company, but various circumstances prevented it from happening for a long time. So, if this had failed, I would have seriously considered giving up on producing stage performances altogether. However, it was a success, which I feel I haven't experienced for a long time, and it went really well. Since this stage adaptation was the initial goal and we've been running at full speed despite uncertainties, I feel like taking a break for now. It's like 'asatte' has been born anew at this moment, so I want to take my time to think about what I want to do next and start from there.

小泉今日子　俳優、歌手　1982年に歌手デビュー。俳優として映画、ドラマなどの主演、出演多数。執筆家としても活躍。2015年、制作会社「明後日」を設立し、舞台、音楽、映画などプロデュース業にも従事。ポッドキャスト番組『ホントのコイズミさん』の書籍シリーズ第3弾『ホントのコイズミさん NARRATIVE』（303BOOKS）が発売。

Actor and singer Kyoko Koizumi debuted as a singer in 1982. She has since starred in numerous films and TV dramas and works as a writer. In 2015, Kyoko founded the production company 'asatte', engaging in producing stage plays, music, and films. Her podcast series 'Honto no Koizumi-san (The Real Koizumi-san)' has been turned into a book series titled *Honto no Koizumi-san NARRATIVE (The Real Koizumi-san NARRATIVE)* published by 303BOOKS.

KEISUKE KAGIWADA Writer

シンプルでスタンダードな服を作り続けてきたアニエスベー。普遍的であるからこそ、それらの服は映画界でも愛されてきた。アニエスベーのアイテムが登場する映画『日曜日が待ち遠しい！』『レザボア・ドッグス』を題材に、ファッションが与える映画への影響についてライターの鍵和田啓介が考察する。

木漏れ日が降り注ぐ南仏の路肩を、1人の女性が足早に闊歩している。膝丈のスカートの裾を閃かせ、ハイヒールでかつかつと石畳を打ち鳴らしながら、職場に続く道を急いでいる。スカートからのぞく長い脚が魅力的な彼女の名は、バルバラ。フランソワ・トリュフォーの遺作、『日曜日が待ち遠しい！』のオープニングシーンだ。いかにもトリュフォーらしい幕開けじゃないか。なんせ彼は超弩級の脚フェチとして知られた監督なんだから。けれどこのシーンから読み取れるのは、彼のフェティシズムの表明だけに留まらない。

殺人容疑で警察から追われる社長の無実を晴らすべく、秘書のバルバラが真相を究明せんと大活躍する本作について、トリュフォーはこう語っている。「快調なリズムに観客を乗せて、映画の進行のしかたに疑問をはさむ余裕がないようにしようと努めました」。確かに本作は、観客に推理する暇を与えず、猛スピードで転換していく。つまり、忙しなく揺れるスカートの裾と軽快な音を刻むハイヒールは、これから始まる物語のリズムを予告してもいるのだ。

劇中に目まぐるしく登場する"くるくる回るアイテム"のイメージが、「快調なリズム」をさらに活気づける。電話のダイヤル、回転扉、床を転がるカメラのレンズ。中でも印象に残るのは、2番目の殺人事件の際、被害者の腕に巻かれた時計に、カメラがズームするシーンだ。画面に大写しにされた腕時計の針は、持ち主が殺されても止まることがない。「まだまだ映画は止まらないぜ！」という宣言かのように。

登場人物をお洒落に彩る道具としてだけじゃなく、台詞で説明されることのない側面を物語るためにも、ファッションアイテムをフル活用する。トリュフォーが世界最高峰の映画作家に数えられる所以がここにある。

そんなトリュフォーを、「俺はファンじゃない」「彼はとても情熱的だけ

'Narrative Fashion' in Film

映画の中の"物語るファッション"

鍵和田啓介

agnès b. has consistently created simple and timeless clothing. Because of their widespread appeal, these clothes have also been loved by the film industry. Taking films like *Vivement dimanche!* and *Reservoir Dogs* that feature agnès b. items as examples, writer Keisuke Kagiwada explores the impact of fashion on cinema.

A woman walks briskly along a roadside in the south of France, where sunlight filters through the trees onto a roadside. Her knee-length skirt flutters as she clicks her high heels on the cobblestones and hurries to her workplace. The woman's name is Barbara, and the long legs that show beneath her skirt are captivating. This is the opening scene of François Truffaut's final film, *Vivement dimanche!* Isn't this the kind of opening we'd expect from Truffaut? After all, Truffaut was a director known for his extreme leg fetish. However, this scene conveys more than just his fetishism.

To clear her boss' name, who is on the run from the police under suspicion of murder, Barbara, his secretary plays a pivotal role in uncovering the truth. Truffaut commented on this film, 'I tried to keep the audience engaged with a lively rhythm, leaving them no room to question the progression of the film'. Indeed, this film doesn't give the audience time to speculate, as it transitions at a rapid pace. In other words, the busy swish of the skirt hem and the brisk tapping of high heels foretell the rhythm of the story that is about to unfold.

The image of 'spinning items' that appear rapidly

throughout the film amplifies the lively rhythm. Telephone dials, revolving doors, a camera lens rolling on the floor. Among these, the most memorable is the scene where the camera zooms in on the watch wrapped around the victim's arm during the second murder. The hands of the wristwatch, displayed prominently on the screen, continue to tick relentlessly even after the owner has been killed. It is as if the film declares, 'The film is far from over'!

Fashion items are utilised not just as stylish accessories for the characters but also to convey aspects of the story that are not explained through dialogue. This is why Truffaut is considered one of the greatest filmmakers in the world.

However, Quentin Tarantino harshly criticises Truffaut, saying, 'I'm not a fan', and 'He is a very passionate but clumsy amateur'. In his movie *Once Upon a Time in Hollywood*, Tarantino continues disparaging Truffaut's films through the narration of Cliff, played by Brad Pitt in the movie adaptation. It's clear that Truffaut's work does not appeal to him. Nonetheless, one can find clear commonalities between the two directors.

ど、不器用なアマチュアだと思う」とこき下ろすのは、クエンティン・タランティーノだ。彼は自ら執筆した小説版『ワンス・アポン・ア・タイム・イン・ハリウッド』（邦題は『その昔、ハリウッドで』）においても、映画版でブラッド・ピットが演じたクリフの語りを通して、トリュフォー映画をディスり倒している。よほどお気に召さないのだろう。にもかかわらず、2人には明らかな共通点が見出せる。

　脚フェチ？確かにそれもある。けれどここで触れたいのは、タランティーノもまた、ファッションアイテムに物語る役割を与えているという点だ。

　彼の初監督作『レザボア・ドッグス』といえば、黒いスーツに身を包んだ6人の男たちがそぞろ歩くクールな冒頭シーンでお馴染みだ。しかし、よく目を凝らすと、彼らのスーツがお揃いじゃないことがわかる。低予算のため、役者がそれぞれ自前で用意したというのが、その理由だ。ミスター・ブロンドを演じたマイケル・マドセンによれば、自分はジャケットとパンツがチグハグだし、ミスター・ピンクを演じたスティーヴ・ブシェミに至っては黒いデニムで参加したらしい。そんな中、かねてから知己のあったあるデザイナーより提供されたスーツを着用しているのが、ミスター・ホワイトを演じたハーヴェイ・カイテルだ。

　この消極的な選択はしかし、不幸中の幸いだったと言えるかもしれない。不揃いなスーツは、やがて仲間割れする6人の個性のバラバラさをも、表現してしまっているのだから。実際、シックで仕立てのいいスーツは、ピンチのときでもコームで髪を撫でつけることだけは忘れないミスター・ホワイトの伊達男ぶりを、強く打ち出すことにひと役買っている。

　冒頭のそぞろ歩きの後、映画は急展開を迎える。観客の目に映るのは、ホワイトが運転する車の後部座席において断末魔の叫びをあげるミスター・オレンジだけだ。何が起こったのかはわからない。わからない

けれど、事態ののっぴきならなさだけは直ちに理解できる。なぜなら、オレンジの白いシャツが鮮血で真っ赤に染まっているから。宝石強盗を決行した彼らが、何者かの密告により警察の返り討ちにあったと判明するのは、もう少し後の話だ。しかも、劇中では肝心の強盗シーンが一切描かれない。つまり、本作の鍵を握る"強盗の失敗"を映像として（つまり、台詞以外で）示すのは、血塗られたシャツだけなのだ。この大胆さには嘆息するしかない。

　6人は最終的に1人残らず息絶える。さらに言えば、タランティーノは次作『パルプ・フィクション』でも、黒いスーツを着たジョン・トラボルタに、あっけない死を与えている。タランティーノ映画において黒いスーツは、これから死にゆく自分たちを予め弔う喪服という意味も担っているのかもしれない。

　タランティーノがいかにトリュフォーを毛嫌いしていようと、劇中のファッションアイテムの使用法において2人が響き合っているのは、もはや明らか。それを鑑みると、『日曜日が待ち遠しい！』の時計も、『レザボア・ドッグス』でカイテルが着用したスーツも（加えて言えば『パルプ・フィクション』でトラボルタが着た黒いスーツも）、同じくアニエスベーのアイテムだったという事実は興味深い。しかし、シネフィルとしても知られる3人だ。もしかするとこれは、映画の神のおぼしめしなのかもしれない。

Fanny Ardant, J-L Trintignant
VIVEMENT DIMANCHE!
UN FILM DE François Truffaut

A foot fetish? That is undoubtedly one aspect they share. But what I want to highlight here is that Tarantino also assigns a storytelling role to fashion items.

His directorial debut, *Reservoir Dogs*, is famous for its iconic opening scene of six men strolling together, all dressed in black suits. However, upon closer inspection, you would notice that their suits are not identical. Due to a low budget for the film production, the actors had to wear their own suits. Michael Madsen, who played Mr. Blonde, mentioned that his jacket and trousers were mismatched, and Steve Buscemi, who played Mr. Pink, wore black denim. Among them, Harvey Keitel, who played Mr. White, wore a suit provided by his designer friend.

This unintended choice, however, turned out to be a blessing in disguise. The mismatched suits inadvertently highlight the distinct personalities of the six men, who eventually turn against each other. The chic and well-tailored suit significantly highlighted Mr. White's dandy persona, who, even in a crisis, never forgets to smooth his hair with a comb.

After the stroll in the opening scene, the film takes a sharp turn. The next thing the audience sees is Mr. Orange screaming in agony in the back seat of Mr. White's car.

We don't know what happened, but we instantly grasp the severity of the situation because Mr. Orange's white shirt is soaked in blood. It's only later that we learn the police ambushed the jewel thieves due to a tip-off. Moreover, the pivotal robbery scene is never shown. The bloodstained shirt is the sole visual clue (in other words, apart from dialogue) depicting the failed heist. This bold storytelling decision is truly breathtaking.

All six men ultimately meet their demise. Furthermore, in his next film, *Pulp Fiction*, Tarantino also dressed John Travolta in a black suit, only to give him an abrupt death. In Tarantino's films, the black suit might be a preemptive funeral attire, foreshadowing the characters' impending deaths.

No matter how much Tarantino may dislike Truffaut, it is evident that the two directors resonate with each other in their use of fashion items in their films. Considering this, it is intriguing to note that the watch in *Vivement dimanche!*, the suit worn by Keitel in *Reservoir Dogs*, and even the black jacket worn by Travolta in *Pulp Fiction* were all items from agnès b. However, all three—Truffaut, Tarantino, and Agnès—are known cinephiles. Perhaps this connection is the work of the cinematic gods.

鍵和田啓介　ライター　雑誌『POPEYE』を筆頭に、様々な媒体にポップカルチャー関連の記事を寄稿する他、映画ポッドキャスト「PARAKEET CINEMA CLASS」のナビゲーターもつとめる。著書にイラストレーターの長場雄との共著『みんなの映画100選』がある。

As a writer, Keisuke Kagiwada writes pop culture-related articles for various websites and *POPEYE* magazine. Keisuke also serves as the presenter for the film podcast 'PARAKEET CINEMA CLASS'. His publications include the co-authored book *Minna-no-Eiga-Hyaku-sen (100 Movies for Everyone)* with illustrator Yu Nagaba.

Seeking an Authentic Self

嘘のない自分を求めて
モトーラ世理奈

Photography: Masumi Ishida
Styling: Takanohvskaya
Hair & Make-up: Eriko Yamaguchi
Interview: Rei Sakai

agnes b. stories

SERENA
MOTOLA

Interview

Interview

海外での仕事や旅を通して感じたのは、人と人の正直な関係と刺激に溢れた出会い。
自分の可能性を広げてくれる新しい居場所を求めて、拠点をロンドンに移す決意をしたモデルのモトーラ世理奈。
新しいステップを踏む彼女の素顔を、かねてから親交のある写真家・石田真澄が撮り下ろした。

モトーラさんは、自分自身を上手にファッションで表現しているなと思います。服がご自身に与えるものってなんですか？

M　強くなれるもの。好きな色やものがあるとテンションが上がります。服を決める上で、私自身の気分を一番大事にしています。人にどう見られるかというより、自分がいい気持ちになるかどうか。気持ちがルンルンしていた方が楽しいから（笑）。

自分らしくあり続けることを支えてくれているわけですね。

M　他から影響を受けることもあるし、常に変化している部分もあるけれど、自分の本当の部分は変わらないでいたい。こう見てほしい、こう見せたいみたいなことが、自分らしくないのかも。

モデルの他にも役者や歌手など、さまざまな表現活動をされていますよね。その中で意識していることはありますか？

M　やることが変わっても、自分自身は変わらないでいること。ロンドンに行っても、このままの感じでいたい。もちろん、文化も環境も違う土地で暮らすことでいろんな影響を受けると思うけれど、自分を装うようなことはしたくない。嘘のない人がかっこいいなと思うから。

ロンドンへの移住はどんな理由で決意したのですか？

M　海外に住みたいというのは、中学の頃から思っていました。それが留学なのか、どんな形かはわからなかったけれど、仕事で海外に行ったり、海外のチームと一緒に撮影をしたりして、もっといろんな人に会ってみたいなって。少し前に初めて訪れたロンドンがすごく楽しくて、直感的に住みたいなと思いました。

中学生の頃に海外を意識したのは、何かきっかけがありましたか？

M　お父さんがアメリカ人で、お母さんが日本人なんですけど、私は日本で生まれ育ちました。自分では日本人だと思っていたんですけど、周りから「ハーフなの？」って聞かれることが多くて。コンプレックスだったわけではないけれど、自分の居場所はここだけではないかもしれないって感じていて。

ロンドンは、感覚的に呼ばれているような、自分の新しい居場所かもしれない。

M　そうですね。あと、私の名前は「世理奈」って書くんですけど、由来があって。私のお母さんは20代のときにアフリカやアメリカに行っていて、私が生まれるときに日本に帰ってきたんですけど、私に「世界を知ってほしい」という願いを込めてこの名前をつけたみたいで。それを幼い頃から聞いていたので、どこかその影響もある気がします。

過去に訪れたロンドンには、どのような楽しさがありましたか？

M　街の雰囲気がよかったのと、友達を通じて知り合った現地の子たちの集まりに行ったらいろんな国の人がいて、若い人のパワーがすごいなと感じました。人との接し方も自分に合うような感覚があって心地よかった。

「人との接し方」というのは、具体的にどんな部分でしょう？

M　もちろん一人ひとり違うけれど、相手の気持ちを考えて喋るところは少し日本人と近いような印象を受けました。一方で、誰にでも心を開くわけではなかったり、自分のことしか考えていないようなところも感じたりして。その正直なところも含めて、

いい意味の適当さがあるというか。日本ではちょっと窮屈に感じるところが、向こうだと何も気にしなくていいやって思えて、自由になれる感じがしました。

お仕事ではどんなことをやってみたいですか？

M　モデルです。みんなでひとつの作品をつくっていく過程が好きで、自分は被写体としてどう携われるのかを考えるのがおもしろい。幼い頃から大好きなファッションの中で表現ができるのが嬉しいですね。もっと違う文化がある場所で、いろんな人とつくってみたいなと思います。

Through her work and travels abroad, Serena Motola has experienced the power of honest relationships and encounters filled with inspiration. Seeking a new environment that would broaden her horizons, she decided to move her base to London. As Serena takes this new step, her true self is captured by long-time friend and photographer Masumi Ishida.

Serena, you express yourself very well through fashion. What does clothing give you?

M　It gives me strength. Wearing colours or items I like lifts my spirits. When I choose clothes, I prioritise my own mood above anything else. It's not about how others see me but whether I feel good in them. It's more fun when I feel cheerful and happy. (Laughs)

So, it helps you stay true to yourself.

M　While I am influenced by others and constantly evolving, I want my true self to remain unchanged. Wanting to be seen in a certain way or trying to present myself in a specific manner might not be true to who I am.

You are involved in various forms of expression, such as acting and singing. Is there something you are mindful of in these activities?

M　No matter what I do, I aim to stay true to myself. Even if I go to London, I want to maintain my sense of self. Of course, living in a place with different cultures and environments will influence me in many ways, but I don't want to put on a false mask. I think people who are honest with themselves are the coolest.

What motivated your decision to move to London?

M　I've wanted to live abroad since middle school. I wondered if it would be through studying abroad or in some other form, but after travelling for work and collaborating with overseas teams, I wanted to meet more people. I visited London for the first time a while ago and had an incredible experience. I felt an intuitive desire to live there.

Was there a specific reason you started thinking about living abroad in middle school?

M　My father is American, and my mother is Japanese, but I was born and raised in Japan. I always considered myself Japanese, but people often asked me, 'Are you half-Japanese?' It wasn't

an inferiority complex, but it made me feel that the place I belong might not be limited to just Japan.

London feels like a place calling to you, a new home, perhaps.

M　Yes, that's right. My name is written as '世理奈' (Se-re-na), and there's a story behind it. My mother travelled to Africa and America in her twenties and returned to Japan when I was born. She gave me this name with the wish for me to 'know the world'. I've heard that story since I was little, so I think that has influenced me as well.

What was enjoyable about your past visits to London?

M　I loved the atmosphere of the city. When I went to gatherings with local people I met through friends, there were people from various countries. I felt a tremendous energy among the young people. The way people interacted with each other felt right for me, and it was very comfortable.

What do you mean by 'the way people interacted with each other'?

M　Of course, everyone is different, but I got the impression that people in London speak considering the other person's feelings, which is similar to Japanese culture. On the other hand, I also sensed that they don't open their hearts to just anyone, and sometimes they only think about themselves. Including that honesty, there's a sort of casualness in a good way. In Japan, some things feel a bit restrictive, but in London, I felt like I didn't have to worry about anything and could feel free.

What kind of work would you like to do?

M　I am a model. I love the process of everyone coming together to create a single piece of work, and I find it interesting to think about how I can contribute as a subject. It makes me happy to express myself in the fashion world, which I have loved since I was a child. I want to create in different cultural settings and collaborate with various people.

モトーラ世理奈　モデル　雑誌『装苑』でモデルデビュー後、数々のファッション誌、広告などで活躍。2018年にはパリコレクションデビューを果たす。同年、映画『少女邂逅』で初の主演をつとめて以降、映画やドラマ、歌手など活動の幅を広げている。2023年6月から渡英し、ロンドンを拠点に活動中。
After making her modelling debut in the magazine *Soen*, Serena Motola has worked in numerous fashion magazines and advertisements, and made her debut at Paris Fashion Week in 2018. Serena moved to the UK in June 2023 and is currently based in London.

Interview

When 'Things I Like' Connect

"好き"がつながる瞬間

立山由紀

アニエスベー ボヤージュ博多大丸店で店長を務める、立山由紀。もともと音楽や映画が大好きだった彼女は、それらのカルチャーを通して自身とアニエス・トゥルブレとの接点を見つけていく。ミュージシャンのパティ・スミスやデヴィッド・ボウイ。さらには、アニエスベーが衣装提供をした映画『レザボア・ドッグス』『パルプ・フィクション』など、アニエスと親交の深いアーティストや親和性のある映画を知るたびに、自身の好きと交差する瞬間があったそう。今回はご自宅にお邪魔し、アニエスベーの宝物の数々を見せてもらった。

Yuki Tateyama is the store manager at agnès b. Voyage Hakata Daimaru. Yuki has always loved music and movies; through these cultural interests, she found connections with agnès b. Musicians like Patti Smith and David Bowie, and films such as *Reservoir Dogs* and *Pulp Fiction*, for which agnès b. provided costumes, offered moments of intersection with her passions. This time, we visited Yuki's home to see her cherished collection of agnès b. treasures.

「デヴィッド・ボウイをはじめYMOなど、世界的なミュージシャンのポートレートを撮影してきた写真家の鋤田正義さんの写真が本当に大好きなんです。彼のドキュメンタリー映画『SUKITA 刻まれたアーティストたちの一瞬』を観ていたときのこと、鋤田さんがニューヨークのハワード通りにあるアニエスベー ギャラリー ブティックを訪れるんです。そこで展示をしていたデヴィッド・ゴドリスとパティ・スミスを写した写真について語り合うシーンがあって。大好きなふたりがつながる瞬間でとても興奮しました。赤いスカートに合わせたゴドリスのTシャツ（P34右上）は日本上陸40周年の節目に合わせて復刻したものです」

アニエスベーが協賛、衣装提供をしてきた映画もチェックをしている立山さん。1994年カンヌ国際映画祭で『パルプ・フィクション』がパルム・ドールを受賞した際のトロフィーと劇中に登場するピストルを手にチャーミングな表情でカメラの前に立つクエンティン・タランティーノ。その時の写真がプリントされたアーティストTシャツ（P34左下）も、立山さんにとって思い入れのある一着。

「旦那さんが記念日にプレゼントしてくれた貴重な一着で、大切にしているあまり、着用することはなくずっと保管しています。最近だと、2022年にアニエスベーが協賛をしたレオス・カラックス監督の『アネット』を観ました。アニエスとは長年にわたって親交を深めてきたフランス人監督ですが、彼の作品を観るのは初めて。ストーリーはもちろんですが、初のミュージカル作品で、劇中歌を担当しているスパークスというデュオの音楽が最高なんです。その後、彼らのアルバムを聞きなおしたり、主演のアダム・ドライバーの出演作品を深掘りしていったり。そうやって、アニエスが発信していることから、自分の好きを見つけていくことも楽しさのひとつ。好きが連鎖したり、気になったものを深掘りしたりしていくとアニエスに行き着くこともあるんです」

天神北にある九州朝日放送の向かいの小さな映画館、KBCシネマ。芸術性に富んだクオリティの高い作品を上映している、立山さんがよく訪れる場所。フューチュラとのコラボレーションのコンビネゾンを着て映画館にも訪れました。

「フューチュラも大好きなアーティストのひとり。鮮やかな色使いや、社会的なメッセージを作品に込める姿勢が好きなんです。このコンビネゾンは彼が制作をする時に着ているものをオマージュしているんです。ちょうどアニエスベー渋谷店がオープンした2019年頃、東京各所でフューチュラ展が開催されていて。渋谷店で1点だけ残っていてゲットできた大事な宝物です」

入社から20年間、バッグや小物などのアクセサリーを提案するボヤージュを担当している立山さん。タイムレスやシンプルをキーワードとするアニエスベーのスピリットは、どのようにアクセサリーへ反映されているのか聞いてみました。

「レザーの質感には独特の癖があって、スタイルや街に溶け込む感じというか。そこがたまらなく良いのです。あとは必ず、アニエスの"好き"が反映されているところ。インスピレーション源はいつも日常や身近なところにある。たとえば、赤や黄色、緑がモザイク状の柄になったバッグがあるんですけど、実はお庭や森、お花の景色だったりするんです。でも結局、アニエスベーらしさって一言では表せない。映画や音楽、アート作品の楽しみ方は人それぞれ自由であるのと同じように、受け取り方もそれぞれ自由でいいと思うんです。それがアニエスベーの面白さですね」

'I absolutely love the photographs by Masayoshi Sukita, who has taken portraits of world-famous musicians like David Bowie and YMO. When I was watching the documentary film *SUKITA*, there was a scene where Sukita visited the agnès b. galerie boutique on Howard Street in New York. He had a conversation with David Godlis, who was exhibiting there, about his photographs of Patti Smith. It was an exhilarating moment to see two of my favourites connect. The Godlis T-shirt I paired with a red skirt (P34 top right) is a reedition released to celebrate the 40th anniversary of agnès b. in Japan.'

Yuki also keeps track of the films for which agnès b. provided costumes and sponsored. One of her cherished Artist T-shirts features a photograph of Quentin Tarantino holding the Palme d'Or trophy and a pistol from *Pulp Fiction* at the 1994 Cannes Film Festival, where the movie won the prestigious award. Tarantino's charming expression in the photo makes this T-shirt (P34 bottom left) a particularly special piece for her.

'My husband gave me this rare T-shirt as an anniversary present, and I love it so much that I've never worn it and kept it safely stored. Recently, I watched *Annette*, a film directed by Leos Carax that was sponsored by agnès b. in 2022. Although Carax is a French director with a long-standing relationship with Agnès, it was my first time watching one of his films. The story was captivating, but what stood out the most was the music by Sparks, a duo who made the songs for Carax's first musical. Afterwards, I revisited their albums and delved into other films starring Adam Driver, the lead actor. This process of discovering new favourites from the things agnès b. promotes is one of the joys I cherish. Sometimes, my interests and explorations lead me back to Agnès.'

KBC Cinema—a small movie theatre located across from Kyushu Asahi Broadcasting in Tenjin-kita—is known for screening high-quality, artistic films. Yuki often visits it. She walks around the cinema area, dressed in a jumpsuit made in collaboration with Futura.

'Futura is also one of my favourite artists. I love his use of vibrant colours and how he incorporates social messages into his work. This jumpsuit is a homage to what he wears when he works. I got it at the Shibuya agnès b. store just when it opened. At that time, Futura exhibitions were all over Tokyo. It was the last piece left in the store, and it's a precious treasure to me.'

Yuki has been in charge of the Voyage line of bags and accessories, for 20 years since she joined the company. We asked her how the spirit of agnès b., which emphasises timelessness and simplicity, are reflected in these accessories.

'The texture of the leather has a unique charm, blending seamlessly with various styles and urban settings. That's what I love about it. Of course, Agnès' personal "likes" are always reflected in the designs. The sources of inspiration are always found in everyday life and close surroundings. For example, there's a bag with a mosaic pattern in red, yellow, and green inspired by garden and forest landscapes or flowers. But in the end, the essence of agnès b. can't be summed up in one word. Just as everyone has their way of enjoying movies, music, and art, the way you perceive agnès b. can be completely personal and unique. That's what makes agnès b. so fascinating.'

立山由紀　アニエスベー ボヤージュ博多大丸店 店長　2003年熊本店にて入社、結婚を機に福岡へ。アニエスベーのバッグや小物を提案するボヤージュを20年間担当。アニエスベーのアイテムでは、仕事服をモチーフにしたトラバーユコレクションが好きで、古着を合わせたオリジナルのスタイリングとチャーミングなカーリーヘアが立山さんのアイコン。

Yuki Tateyama joined agnès b. at the Kumamoto store in 2003 and moved to Fukuoka when she married. For 20 years, Yuki has been overseeing Voyage, agnès b.'s bags and accessories line. Among the agnès b. items, Yuki likes the Travail collection, which is inspired by work clothes. Her original styling often combines vintage clothes, agnès b. items and her charming curly hair.

鋭い洞察力と繊細な表現で、見る人、読む人を惹きつける、お笑い芸人、作家の又吉直樹。
服を選ぶ時の眼差し、素材や柄、サイズ感を自由に組み合わせた着こなしを見ていると、
コント作りや執筆活動と同じように、自身の表現方法のひとつとして大事にしていることがうかがえる。
服を好きになった原体験や、今のスタイルに行き着いた経緯。沸々と湧き出る、服と創作の話。
アニエスベーと私物を合わせた、セルフスタイリングも披露してもらった。

遊びは、創作
又吉直樹（芸人, 作家）

Playing and Creating
NAOKI MATAYOSHI (Comedian, Writer)

Photography: Yuto Kudo
Hair & Make-up: kika
Text: Megumi Koyama

アイテム選びやスタイリングもあっという間でしたが、直感なのでしょうか？

M　小学生の頃からサッカーをやっていたので、毎日ジャージばかり着ていました。世界各国のクラブチームのユニフォームを雑誌で見ていて、買えなくてもこれが欲しいなと思ってよく眺めたりしていたんです。ユニフォームの色や、パンツは何を合わせているんだろう？と考えたり。そうやって色の合わせ方を自然と学んでいたような気がします。

スポーツウエアから服に興味を持ったきっかけは？

M　おばあちゃん家に行く時だけは「ジャージで行ったらあかん」って言われていて。近所にある駅前のデパートに服を買いに行って、自分でブラックジーンズを選びました。偶然、おばあちゃんの家に行った帰りに友人に誘われて、ベースボールシャツとブラックジーンズで遊びに行ったんです。そしたら、クラスの女子の団体に遭遇して「又吉がジャージ着てない！めっちゃオシャレやん。そっちの方が絶対いいよ」と言われて、その日からですね（笑）。

そこからファッションに目覚めた？

M　姉に「テレビで見たアメリカ村っていうところ行ってみた

い」って相談したんです。姉がGジャンを腰に巻いていたら、店員さんに「Gジャンもいいけど、ネルシャツの方がええで」って言われてて。ネルシャツって何！？ってなったのを覚えています。古着屋をめぐって、自分でこれだと思うネルシャツを買ったんです。その日購入したネルシャツと紺色のワークパンツを合わせて、みんなの前に現れた時には、革命が起きましたね（笑）。そこからもう、服がめっちゃ好きになりました。

中学生時代は、どのようにファッションを楽しんでいましたか？

M　アメ村への熱は少し落ち着いていて、両親のクローゼットで眠っている服を全部出して、メンズやウィメンズ関係なく着てみたり。Aラインのコートを引っ張ってきて、下に広がっているから、ストレートのパンツよりも、フレアの方が良いな、じゃあ今度ベルボトムのパンツ買ってみよう。みたいなことをやっていましたね（笑）。元々ある洋服を改造したり、切って布を張り合わせたり、ストレートパンツにジッパーをつけてみたり、自分でDIYすることにも興味を持つようになりました。

着るだけではない、服の楽しさというものに早くも気づいていたんですね。ご自身のスタイルはどのように出来上がっていったのでしょう。

M　誰にも教わったことはないですし、誰かの真似をしたこ

又吉直樹 アニエスベーを着る。

agnès b.

No. 7

ともないんです。アメ村の三角公園や、心斎橋駅の前にずっといると、おしゃれなお兄さんやお姉さんがいるので、やばいなと思う人がいると、後をつけて行く。そういう人に着いていくと、雑居ビルの4階とかに辿り着いたりして（笑）。「君何しにきたの？」って話しかけられて、色々お店を教えてもらったり。観察をつづけていると、この人、文化的な匂いがするな、音楽をしている人はこういう格好しているんや、悪そうな人はこういう格好やな、ペインターパンツとか履いて、大きめのTシャツを着ている人はダンスやろうなとか。その人たちの主義主張や、スタイル、服を観察していく中で、じゃあ自分はなんやろう？ってよく考えていたと思います。自分の考えや、やることを辿っていくと、中学3年生の時には、蛇柄のシャツを着て、伊達眼鏡、ハンチングを被って、革靴を履いていました（笑）。

今日は、アニエスベーのアイテムでスタイリングをしていただきました。どのようなストーリーがあるのでしょうか？

M　モノクロの柄のスウェットパンツは、おもしろいなと思って最初に手に取りました。シンプルにTシャツを合わせる方が、スタイリングも組みやすいような気がしているんですけど。ゼブラ柄のブルゾンとは、色は合っているけど柄と素材が違う。それがどういう風に作用するかなって思ったんです（P38）。ワークウエアのデザインも好きで、よく着ているアイテム。生地がしっかりしていて、きちんとしたお店にも着ていけるような清潔感が良いなと思いました。だからこそ、カチッとさせすぎず、渦まき柄のシャツ合わせがラフで良いかなって（P37）。映画『KILL BILL』の古着のTシャツは、黄色の相乗効果を狙ったスタイリングです。黄色と合わせることで、オレンジのカーゴパンツが引き立っていますよね（P41）。

創作においてもいい塩梅ってあると思います。又吉さんの考える、いい塩梅とは何でしょう？

M　いい塩梅にも、時代性みたいなものがあると思うんです。江戸時代とかで言ったら、町で暮らす人たちはあんまり派手な格好をしたらあかんから、裏地めっちゃおしゃれやったり。時代によって服のバランス感とか、まとめ方って変わってくると思うんです。僕の場合は、その時代

で一番気持ちいいとされているところから、ちょいずらしですね。ちょいずらしがイケている時代やったら、さらにちょいずらし。みんなより、ちょっと踏み外していたいっていうのは、いつもあります。それは僕自身が、平均的な人間じゃないというのもあるし、自分のスタイルの表明でもあるんです。芸人って舞台立った時に、おしゃれとかダサいとか関係なく、なんとなくこの人がボケなんかなとか、ツッコミなんかなとか、わかるじゃないですか。裏切られる時もあるんですけど、みんな自分のスタイルで表現していると思うんですよね。

創作のインスピレーションは、どんなところから生まれるのでしょうか？

M　毎日、何かしら作るものがあるので、完全な休み、みたいなものはないんです。だからほぼ24時間営業で、いついかなる時も遊びつつ、何かしら、考えていますね。ひとりでお酒を飲んでいても、飲みながら、明日の自分の助けになることを、これは仕事じゃないと思いながらずっと考えています。ちょっとせこいんですけど、人とお酒を飲む時も、相手の相談を聞きつつ、自分の仕事のクリエイティブの方に持っていく。ベタなので言うと「無人島にひとつだけ持っていくモノの、100番目ぐらいって何やと思う？」とか。みんな疑うんですよね、仕事してないか？って（笑）。みんなの話を聞きながら、なるほどと思いつつ。その質問が自分に回ってきた時に答えていると、ディスカッションになる。それで、一本エッセイが書けるとか、コントになることもありますね。

人との会話の中で、生まれることが多いんですね。

M　よく思いついたなということも、人と喋ってる時に、ポンって出たりもするので。それを元に作り直すときに、もっと良いものや、新しいものが生まれたりもする。映画やファッション、音楽でもそうなんですけど、文学をやる時に、文学から影響を受けると直接的じゃないですか。自分が思いつかないようなものを、思いつきたい。でも、自分が思いつくようなものは、自分が思いつくものだから、ひとりで考えているとそのジレンマに陥ってしまうんです。外部的な要因を自分にどんどん当てていくというのは、積極的にやっているかもしれないですね。

With his sharp insight and delicate expression, comedian and writer Naoki Matayoshi captivates his viewers and readers. When you observe how Naoki selects his clothes, combining materials, patterns, and sizes freely, you can see that—just like in his skits and writing—fashion is an important form of self-expression for him. Naoki shares the experiences that made him fall in love with clothes, his journey to his current style, and his bubbling thoughts on fashion and creativity. Naoki also showcased his self-styling, combining agnès b. pieces with his own items.

You quickly chose the items and styled your outfit today—did you do it by intuition?

M   Since I played football in elementary school, I only wore sweatpants every day. I often looked at magazines featuring team uniforms from around the world, thinking about how much I wanted them even if I couldn't buy them. I would think about the colours of the uniforms and what pants they were paired with. I feel like I naturally learned how to coordinate colours that way.

What sparked your interest in fashion from sportswear?

M   I was told that it wasn't polite to wear sweatpants to visit my grandma's house. By chance, I was wearing a baseball shirt with black jeans one day to visit my grandma, and after that, a friend invited me to meet up. We ran into a group of girls from my class, who said, 'Naoki isn't wearing sweatpants! You look so stylish. This is much better.' That's when it all started. (Laughs)

Did that spark your interest in fashion?

M   I asked my sister to take me to America-Mura, a vintage store I saw on TV. At the store, when my sister wrapped a denim jacket around her waist, the shop assistant told her, 'The denim jacket is nice, but a flannel shirt would be better.' I didn't know what a flannel shirt was, but we went to different vintage shops, and I found the perfect flannel shirt. When I paired it with navy work pants and showed up in front of everyone, it was a revolution. (Laughs) From that day, I became passionate about fashion.

How did you enjoy fashion when you were in your junior high school years?

M   My enthusiasm for America-Mura had settled a bit, and I started pulling out all the clothes from my parents' closet, trying on men's and women's clothes. I would take an A-line coat, and because it flared out at the bottom, I thought flared pants would be better than straight ones, so I decided to buy bell-bottom pants after that. I was just doing things like that. (Laughs) I also became interested in DIY fashion, modifying existing clothes, cutting and stitching fabrics, and adding zippers to straight pants.

You noticed quite early on that the joy of clothing goes beyond just wearing it. How did you develop your style?

M   I never learned from anyone or tried to imitate anyone. I spent a lot of time hanging out at America-Mura in front of Shinsaibashi Station, where stylish people are. If I saw someone who looked fashionable, I'd follow them to some old building, and they would recommend me different shops to go to. By observing their principles and beliefs, styles, and clothes, I often thought about my identity. Tracing my thoughts and actions, by the time I was in my third year of junior high school, I was wearing a snakeskin shirt, fake glasses, a hunting cap, and leather shoes. (Laughs)

You've styled agnès b. items today. What's the story behind your choices?

M   I first picked up the monochrome patterned sweatpants because I thought they looked interesting. Pairing them simply with a T-shirt is an easy way to create a cohesive look. The zebra-patterned blouson matched in colour but had different patterns and materials (P38). I was curious about how this combination would work. I also like workwear designs and often wear items like this. The fabric is sturdy and has a clean look that is suitable for more formal settings, which I like. That's why pairing it with a patterned shirt for a more relaxed feel looks good (P37). The vintage *Kill Bill* T-shirt was chosen to play up the yellow tones. Pairing it with a yellow T-shirt enhances the orange cargo pants, making them stand out nicely (P41).

In creation, there is something that we call a good balance. What do you consider to be a good balance?

M   A good balance has a certain sense of the times. For example, in the Edo period, townspeople couldn't dress in a flashy style, so they would make the linings of their clothes very stylish. In my case, it's about shifting slightly from what's considered most comfortable in that era. If slight deviations are popular at that time, then I'll shift further. I always want to step a little out of the norm because I'm not an average person, and it's a way of expressing my style. When a comedian steps onto the stage, you can often tell who's the funny one or who's the serious one, regardless of their fashion. Sometimes you're surprised, but usually, everyone expresses themselves in their style.

Where does your creative inspiration come from?

M   I'm almost always operating 24 hours a day, thinking and playing simultaneously. Even when I'm drinking alone, I'm always contemplating things that will help me tomorrow, thinking it's not work, but it is. It's a bit sneaky, but even when I'm drinking with others and listening to their problems, I steer the conversation towards my creative work. For example, asking something cliché like, 'What do you think the 100th item you'd take to a deserted island would be?' I listen to everyone's stories. When the question comes back to me, I answer, and it becomes a discussion. Sometimes, I end up writing an essay or creating a skit based on that conversation.

It seems that many of your ideas come from conversations with others.

M   Sometimes, I come up with ideas while talking to people; sometimes, they just pop into my head. When I rework these ideas, sometimes they evolve into something even better or entirely new. It's similar with movies, fashion, and music. When working on literature, being influenced directly by other literature can be too straightforward. I want to come up with ideas I wouldn't usually think of. But if I'm alone, I tend to think of things that I would naturally come up with, which creates a dilemma. Actively exposing myself to external influences is something I do deliberately.

又吉直樹　芸人、作家　2003年に綾部祐二と「ピース」を結成。現在は、執筆活動にくわえ、テレビやラジオ出演、YouTubeチャンネル『渦』での動画配信など多岐にわたって活躍。著書に、小説『火花』『劇場』『人間』、エッセイ集『第2図書係補佐』『東京百景』。2023年に10年ぶりのエッセイ集『月と散文』を発表。

Naoki Matayoshi formed the comedy duo 'Peace' with Yuji Ayabe in 2003. In addition to his writing activities, Naoki appears on television and radio and is active on his YouTube channel, 'Uzu'. His published works include the novels *Hibana (Spark), Gekijo (Theatre),* and *Ningen (Human),* as well as essay collections *Dai Ni Toshokari Hosa (Assistant Librarian No. 2)* and *Tokyo Hyakkei (Hundred Sceneries of Tokyo).* In 2023, Naoki released his first essay collection in 10 years, *Tsuki to Sanbun (The Moon and Prose).*

ブティックで流れる音楽からラジオまで、アニエスベーは長年に渡り、独自のサウンドアイデンティティを築き上げてきた。アニエスベーで20年以上働くヤン・ル・マレックはその立役者といっても過言ではない。好奇心旺盛で何よりも音楽を愛するヤンのおかげで、アニエスベーの"音楽"は確立されてきた。サウンド・アイデンティティ・キュレーターという珍しい肩書きを持つヤンに、アニエスベーと音楽にまつわる特別な思い出を教えてもらった。

Let's Share Music

音楽を分かち合うために
ヤン・ル・マレック

　CDが登場する前は、カセットテープでラジオ番組を録音して好きな音楽を聴いたり、レコードも集めていました。12歳か13歳の頃です。当時はインターネットもなかったので、可能な限りレコードショップに足を運びました。コレクションを増やすという目的ではなく、お気に入りの音楽をいつでも聴けるようにしたくて。僕の父は船乗りで、80年代のはじめに日本からパイオニア製のターンテーブルを持ち帰ってきてくれたというのも思い出深い。現在、33回転のLP盤は4000枚以上、45回転のEP盤とCDはもう数え切れないくらいたくさん持っています。

　2003年、店舗の内装を担当する部署で数年働いた後、僕はアニエス・トゥルブレにお店で流れる音楽の選曲を担当させて欲しいとお願いしました。当時も今もサウンド・アイデンティティという部署があるブランドはほとんどなくて、素晴らしいアイデアだと感じました。アニエス・トゥルブレが知っているミュージシャンとの架け橋になり、彼らの音楽をブティックで流す。まずは、店舗にレコードのリスニングステーションを設け、同じレーベルから5枚のアルバムを選定。それらを流してお客様に紹介するところからはじめました。一定期間リスニングステーションで取り上げられたレコードはお店に保管してBGMとしても流すというシステムでした。

　2018年9月にはウェブラジオも開設。実は2016年末から開設に向けて準備を進めていたのですが、選曲にアニエスベーらしさと一貫性を見出すのは大仕事だと感じていました。開設当初はアニエスベーの活動にリンクする14のテーマに基づいてアルバムを選定。その中から、5000曲を超える音楽を1曲ずつ選曲していきました。今では、万枚のアルバムが取り上げられています。

　僕は音楽を聴くときにストリーミングを使いません。レーベルのカタログをリサーチしてコンタクトを取ります。アーティストから直接楽曲を

購入し、ラジオで流す。実際に人に会って話をすることで、音楽の世界で何が起こっているかを把握し、ミュージシャンやレーベルの情報をインプットする。常に勉強を続けていきたいと思っています。

　アニエスベーは、コンサートやフェスティバルへの協賛も行っています。数年前からサポートを続けているフランスの実験音楽フェスティバル、「ソニック・プロテスト」に関しては、そのサポートを継続するために特別な企画を考えたりしました。例えば、プロデューサー、ソングライター、カルト・ミュージシャンであるキム・フォーリーにアニエスベーとラフ・トレードが共同で企画したライブに出演してもらい、ブティックでのサイン会用に限定版のスクリーンプリント・ポスター（現在ではコレクターズ・アイテム）を制作、販売することで協賛金を集めたりもしました。

　僕たちはアーティストだけでなく、インディペンデント・レーベルも支援しています。2011年にフランスではじめて開催されたレコード・ストア・デイズの期間中に僕たちはラフ・トレードのフランスへの再上陸を支援しました。その後、ラフ・トレードとパートナーシップを結び、6年間にわたり、ブティックで80回以上のライブを開催。ラフ・トレードだけでなく、パン・ヨーロピアン・レコーディングとも協業することで、新しいアーティストの発掘も行ってきました。

　アニエス・トゥルブレには、自分が信じるアイデアに「イエス」と言い、とことん突き進む素晴らしい才能があります。アニエスベーには、エコロジーやチャリティ、そして強い倫理観を持ったインディペンデントなクリエイションをサポートするという、僕が大好きで強みにもしている姿勢がある。これからもアニエスベーはインディペンデントアーティストを支援し、応援していきたい。音楽は分かち合うためにあるのだから。

アニエスベーがフランスの実験音楽フェスティバル、「ソニック・プロテスト」と共同で制作した、キム・フォーリーのラストツアー『Mudhouse Tour 2012』のポスター。

Poster produced by agnès b. in partnership with the Sonic Protest Festival to support the last tour of Kim Fowley, *Mudhouse Tour 2012*.

1998年、フランスのバンドAir のファーストアルバム『Moon Safari』の宣伝としてアニエスベーが無料で配布したレコード。後に彼らの代表曲となる『Sexy Boy』が収録されている。当時、アニエスベーは彼らに衣装も提供していた。

A 45rpm flexi disc distributed for free by agnès b. in 1998, to promote Air's debut album *Moon Safari*, which includes their signature song 'Sexy Boy'. At that time, agnès b. also dressed and supported them.

ラフ・トレードとアニエスベーがパン・ヨーロビアン・レコーディングの15周年を記念してリリースしたEP盤。パン・ヨーロビアン・レコーディングのスターでありアニエス・トゥルブレの知人でもあるクドラムとハナ・ワシムのエクスクルーシブトラックを収録。

An EP released by Pan European Recording, Rough Trade, and agnès b. to celebrate the 15th anniversary of Pan European Recording. It features exclusive tracks by Koudlam (a star of the label and an acquaintance of Agnès) and Hanaa Ouassim.

From the music played in the boutiques to radio shows, agnès b. has developed a unique sound identity over the years. Yann Le Marec has worked at agnès b. for over 20 years. Thanks to his curiosity and the importance he places on discovering rare and exclusive music, Yann has helped define this identity. We asked Yann, who holds the rare title of Sound Identity Curator, to share his special memories with agnès b. and music.

Before the advent of CDs, I used to record the radio and my favourite programmes on tape so I could find the tracks I liked; I also collected records. I was 12 or 13. Back then there was no internet, so I used to go to record shops whenever I could. The idea wasn't to build up a collection, but simply to be able to listen to my favourites. My father was a sailor and in the early 80s he brought back a Pioneer turntable from Japan, which was great because it was the first one with variable speed control. Today, I have over 4,000 33rpm, I don't know exactly how many 45rpm and CDs.

In 2003, after a few years working in the decoration and image department for the stores, Agnès asked me to take charge of selecting the music played in the shops. It was, and still is, a rare thing: very few companies have a full-time person in charge of their sound identity, and I thought it was a brilliant idea! The goal was to build bridges between the musicians that Agnès knew and dressed to promote their music in the boutiques. In the beginning, we introduced listening stations that showcased five albums from the same label, to promote them and make them known by our customers. It was a system where after a certain period, the shops could keep the records and play them as the shop BGM.

The web radio was launched in September 2018. I started working on it at the end of 2016. Finding unity and coherence in the choice of the titles is quite a long job! I first created an initial playback scenario, building up a disc library based on 14 themes linked to our musical activities and Agnès' tastes. When we launched it, there were over 5,000 tracks, selected one by one. The web radio now has more than 10,000 titles!

I don't use any streaming platform to listen to music. I do my research via the catalogues of the labels I then contact. We buy tracks directly from the artists to play them on the radio. I find out what's going on in the music world by meeting people and talking to them, and I note down information about musicians and labels. You can always learn from others and you'll never know all the music out there.

agnès b. also supports concerts and festivals. To continue supporting the Sonic Protest festival (an experimental music festival in Paris), which we had been supporting for several years, we organised a special event. We arranged for producer, songwriter and cult musician Kim Fowley to appear during the *Rough Trade on Tour chez agnès b*. We produced limited edition screen-printed posters (now collector's items) for a signing session in the boutique and sold them to raise funds for the festival.

In addition to supporting artists, we also support independent labels. During the first Record Store Day in France in 2011, we helped Rough Trade re-enter the French market. Over the next six years, we hosted more than 80 in-store concerts in partnership with Rough Trade. We have also collaborated with Pan European Recording, to foster long-term partnerships to discover and promote new artists.

Agnès has this incredible ability to say yes to the ideas she believes in and to go all the way. There is a commitment at agnès b. that I love and that is a strength: supporting ecology and charities, as well as independent creation with strong ethics. As an independent brand, agnès b. supports and encourages independent artists. It's unique, and I'm proud to have been part of it for many years, and to share this state of mind through music: music is all about sharing.

ヤン・ル・マレック　ウェブラジオ＆サウンド・アイデンティティ・キュレーター　パリにて店舗の内装を担当する部署に配属された後、2003年にアニエスベーの世界観を音楽で表現する業務に従事。店内で流れる音楽の選曲や、アニエスベーラジオの立ち上げにも携わる。音楽を通して、アニエスベーのクリエイションを支えている。

After being assigned to the department in charge of shop interiors in Paris, Yann Le Marec began working on expressing the world of agnès b. through music in 2003. Yann was also involved in selecting th

Our Motto:
'Since We Have the Chance'

合言葉は「せっかくだから」
村上 萌

MOE
MURAKAMI

Photography: Asuka Ito
Text: Yuki Namba
Special thanks to Ken Tokura

agnès b. stories

Interview

Interview

東京と長崎の二拠点生活を送る村上 萌。彼女の自宅は、長崎市内から少し離れた静かな街にある。
プロサッカー選手である夫の都倉 賢の移籍に合わせて訪れた、それまで縁もゆかりも無かった長崎という土地。
自然豊かなこの場所での暮らしが、子どもの成長や家族の在り方に与えるポジティブなエネルギーとは。

快い風が吹き抜ける、見晴らしのいい丘の上に建つ一軒家。玄関先の階段を上ると、お手製のブランコに乗った娘の杏ちゃんが、満面の笑みで私たちを出迎えてくれた。暖かな日差しが入る茶の間にはゆっくりと時間が流れ、家族の笑い声が響き合う。

「娘がいま5歳なんですが、思考に対して言葉が追いつくようになったのは長崎へ来てからなんです。毎日、どうしてこうなっているの？と疑問が多い中で、その答えというのは自然の中にある気がしていて。家の近所に飲食店があまりないので、庭でハーブを育てたり、地元の漁師さんと一緒に海に魚を獲りに行ってみたり、そういう五感を通した、原体験になるような経験がすごく多いんです」

村上さん自身も、過去に自然から得た貴重な原体験があったそうだ。

「幼い頃、森の中にある学校に通っていたんです。自然の表情って毎日違うし、同じ葉っぱも木の実もない。自然の中での体験って隣の子と答えが違うのがすごく面白くて。テストだと明確に点数でスペック分けされてしまうけれど、自然の中では上も下もない。その体験が多いほど、この先比較されることがあっても自信を無くすことはないと思うんです。娘にはそういった自信を持てる経験をさせてあげたいなと思っています」

村上さん家族が、日常の出来事や感情を記録している「THE FAMILY NOTE」。いま過ごしている日々は子どもの成長や環境の変化とともにいつかは終わってしまう。二度と繰り返すことのない特別な感情や体験をノートに記録することで、その時の感情の受け皿にもなり、家族みんなにとっていい影響があるそう。アスリートである夫の都倉さんは、スケートボードを習っている杏ちゃんの姿から大きな影響を受けていると話す。

「僕はサッカーをやってきた中で成功体験はありましたが、チームメイトもいるので本当に自分だけの成功なのか検証しづらい部分があって。スケボーって失敗イコール転ぶことで、成長の過程において痛みを伴うので、『No pain, No gain』の原則を身体で体験できるものだと思うんです。以前娘がスランプに陥ってしまい、2ヶ月ぐらいドロップインが飛べない時期があったんですね。でも急にまたそれができるようになって、そこからいままで以上に楽しそうにしていて。初めて会った方にもよくその話をしているので、本人の自信にもなっているんだと思います。そんな様子を見ていると、これが成長の本質だなと思うし、すごくインスパイアされますね」

親から子どもに与えられる原体験。そして、それを体験した子どもの成長する姿が親にもたらす新しい発見。互いに反応し合う家族の在り方は、東京とは違ったスピードで時間が流れる土地だからこそ、確かに感じられるものなのかもしれない。都倉さんの転勤や自身のキャリア、子どもの成長に伴い変化していく日常の中で、長崎という土地でしか得られない毎日を大切に過ごしている村上さん。さまざまな環境の変化を、自分なりに楽しむための秘訣を教えてくれた。

「『せっかくだから』と思うようにしています。どうしたら今日より明日が楽しくなるかなって考えてみれば、絶対ひとつぐらいは改善案ってあると思うんです。せっかくだからそれを試しにやってみれば、毎日がちょっとずつ楽しくなるかなって。花瓶に花をいけてみたり、風鈴を飾ってみたり、簡単なことでいいのでひとつやってみると、『あ、こんなことか』って動きだしたりする。せっかくだから精神で、小さなことから始めてみるのが大切なんだと思います」

The home of Moe Murakami—who divides her time between Tokyo and Nagasaki—is located in a quiet town within a short distance from Nagasaki City. They moved here to accommodate her husband, Ken Tokura, a professional soccer player, who had transferred to another team in Nagasaki despite having no connection here. How does the lifestyle in this nature-rich place positively influence their children's growth and the family's well-being?

A house stands on a hill with pleasant breezes and a clear view. On the front steps, their daughter An, swinging on a handmade swing, greets their guests with a beaming smile. The home of Moe Murakami, who divides her time between Tokyo and Nagasaki, is located in a quiet town within a short distance from Nagasaki City. The living room, filled with warm sunlight, has a leisurely flow of time, resonating with the laughter of the family.

'My daughter is five years old now, and it was after we moved to Nagasaki that her words began to catch up with her thoughts. Every day, she asks, "Why is this like this?" and I feel that many of the answers lie in nature. Since there aren't many restaurants near our home, we grow herbs in the garden and go fishing with local fishermen. These kinds of sensory experiences form a lot of fundamental memories for her.'

Moe herself also had valuable experiences with nature in the past.

'When I was a child, I attended a school in the forest. Since the expressions of nature are different every day, and there are no identical leaves or berries, the experiences in nature vary from one child to another, which I found fascinating. At school, we are categorised by scores, but in nature, there is no ranking. The more such experiences you have, the less likely you will lose confidence when you face comparisons later in life. I want my daughter to have these confidence-building experiences as well.'

The Murakami family records their daily events and emotions in 'The Family Note'. The days they are experiencing now, along with their children's growth and changes in their environment, will eventually end. Recording these unique and unrepeatable emotions and experiences in a notebook serves as an outlet for those feelings and has a positive impact on the entire family. Ken, her husband and an athlete, mentions that he is greatly influenced by watching their daughter, An learn to skateboard.

'I've had successful experiences in soccer, but since it's a team sport, it's hard to determine if the success is my own. With skateboarding, falling is failing, and the growth process involves pain, so it's an activity where you can physically experience the "no pain, no gain" principle. There was a time when our daughter hit a slump and couldn't do a drop-in for about two months. But then, she suddenly managed to do it again and seemed to enjoy it even more. I often share this story with people we meet for the first time, which has boosted her confidence. Watching her, I feel that this is the essence of growth, and it greatly inspires me.'

The experiences parents provide their children, and the discoveries that come to parents from watching their children grow as they undergo these experiences. The way a family interacts and responds to each other can be genuinely felt in a place where time flows at a different pace than in Tokyo.

Amid daily life that changes with her husband's job transfers, her career, and her child's growth, Moe cherishes the unique daily experiences that can only be found in Nagasaki. She shared her secrets for enjoying the various changes in her environment in her way.

'I try to adopt a "since we have the chance" mindset. If you think about how to make tomorrow more enjoyable than today, I believe there's always at least one way to improve it. Since we have the chance, why not try it? Doing so can make each day a bit more enjoyable. It can be something simple, like putting flowers in a vase or hanging a wind chime. When you try something small, you might think, "Oh, it's this simple", and start moving forward. I believe it's important to start with small things with a "since we have the chance" attitude.'

村上萌 ライフスタイルプロデューサー 「季節の楽しみと小さな工夫」をコンセプトにウェブサイトの運営を始め、連動した雑誌の刊行や週末イベント、ECストアの運営、その他空間や商品などのプロデュースを手がける。出身地である横浜市内に1000坪の庭「COMMON FIELD」をプロデュースし、敷地内にて「GARTEN COFFEE」を運営。

With the concept of 'seasonal enjoyment and small innovations', Moe Murakami runs a website, publishes a related magazine, hosts weekend events, manages an e-commerce store, and produces spaces and products. In her hometown of Yokohama, Moe produced a 3,300-meter-square garden called COMMON FIELD and operates 'GARTEN COFFEE' within the site.

写真家の服部恭平がアニエスベーのTシャツを着た友人を写した一枚の写真が、パリのアニエスベーに飾られたことがあった。それもアニエス・トゥルブレ本人が気に入ってくれたことから。彼が写真家を目指すと決めた時に憧れていた場所が、青山にあるアニエスベー ギャラリー ブティック。憧れの場所に通っているうちに、いつしか仕事をするようになって、さらにはパリの店舗で自分の写真が飾られるなんて、そんな夢のようなことがおこったのだ。自身と写真との関係について、撮り下ろしの写真とともに話を聞いた。

Finding Your Identity in Everyday Life

日常の中に見つけた、自分らしさ
服部恭平

写真をはじめるにあたって、フィルムカメラを選んだのはなぜだったのでしょうか?

H　プロダクトとしての「フィルム」が好きなんです。写真を仕上げていく工程でフィルムという物とフィジカルに触れる時間が魅力的だと思っています。今回撮り下ろした写真に写っているパズルや砂時計もそうで、なにか新機能が付いているとかではなく普遍的な物。そういったシンプルでずっと受け継がれてきた物に惹かれます。

アニエスベーとの出会いのきっかけは?

H　写真家として活動し始めた時、青山にあるアニエスベー ギャラリー ブティックに憧れていたんです。ワールドワイドでありながら、有名無名問わずアーティストを支援するスタンスがかっこよくて、いつかここで展示したいと思いよく足を運んでいました。ギャラリーや店舗に通っているうちに、縁があってお声がけいただき、アニエスベーのTシャツを着用した友人を撮影することになって。そこで撮影した写真をアニエス・トゥルブレさんが気に入ってくれて、パリの店舗に飾ってもらえたことは本当にうれしかったですね。

今回撮り下ろしていただいた写真は、お父様や家、海など。日常的なものが被写体になっています。どんなストーリーがあるのでしょうか?

H　東京の自宅と大阪の実家、そして家族旅行で訪れた淡路島で撮影しました。子どもの頃からよく家の車で遠出していたんです。いつも父が運転をしてくれていたんですが、大人になった今、僕自身もあの時の父のように運転するのが大好きで「僕が絶対運転する」って。ふと助手席に座っている父を見たとき、感慨深くなってシャッターを切りました。離れて暮らしてから、両親が年を重ねていくことをより実感するようになって。それが決して嫌とかではなくて、まだ自分にはない感覚だからとても惹きつけられますね。あと、実家にちょうどボタンが外れてしまったアニエスベーの白いシャツを持っていって、母にそのボタンをつけてもらいました(笑)。その時あったピンクッションにアニエスベーのアイコンでもある星モチーフのピンバッジを刺してみたり。自宅の中でも光や色が綺麗な瞬間があったらパッとカメラを向ける。そうやっていつも生活の延長線で撮影しています。

友人や家族、自宅の中で見つけた美しい瞬間など、服部さんの写真からは私的なストーリーを感じます。

H　心地よいと感じる暮らしや、生きていく上で出逢ったものを写真におさめたいという思いがあります。僕が語りたいのは強いメッセージではなく、自分が美しいと思うものや、どんな人間なのかを写真を通して伝えたい。それが自分との関係性において距離が近いものになる程、自分だけの表現になっていく気がしているんです。

A photograph taken by photographer Kyohei Hattori, featuring a friend wearing an agnès b. T-shirt, was once displayed at an agnès b. store in Paris. This happened because Agnès Troublé liked the photograph. When Kyohei decided to become a photographer, the agnès b. galerie boutique in Aoyama, Tokyo, was a place that he admired. While Kyohei frequented the shop, he eventually began to work for them, and then the unbelievable happened—his work was displayed at the store in Paris. Alongside his new photographs, we spoke with Kyohei about his relationship with photography.

Why did you choose analogue film when you started photography?

H   I love film as a product. The process of physically handling film during the development of photographs is very appealing. It's the same with the puzzles and hourglasses featured in the photos I took this time—they are timeless objects that are common and don't bear any new features. I am drawn to these simple, enduring items that have been passed down through generations.

How did you first discover agnès b.?

H   I admired the agnès b. galerie boutique in Aoyama since I started working as a photographer. I was drawn to their worldwide presence and cool stance of supporting famous and unknown artists. I often visited, hoping to exhibit my work someday there. As I frequented the gallery and store, I had the opportunity to connect with them and was invited to photograph a friend wearing an agnès b. T-shirt. Agnès Troublé liked the photo, and it was displayed in a store in Paris. That was a truly happy moment for me.

The photos you took this time feature your father, home, the sea, and other everyday subjects. What stories do these images tell?

H   I took these photos at my home in Tokyo, my family home in Osaka, and on a family trip to Awaji Island.

Growing up, we often went on long drives in our family car, with my father always at the wheel. Now that I'm an adult, I've developed a love for driving, just like my father, and I always insist on driving. There was a moment, while my father was sitting in the passenger seat, when I felt a deep nostalgia and decided to capture it with my camera. Living apart from my parents has made me more aware of them growing older. It's not something that bothers me; instead, it's a feeling I don't fully understand yet, and that fascinates me. Also, I brought along with me a white agnès b. shirt with a missing button to my family home so I could ask my mother to sew the button back on. (Laughs) I found a pin cushion with a star-shaped pin badge, which is an icon of agnès b., and used it for fun. Even at home, I'd quickly grab my camera when I see a moment with beautiful light or colours. That's how I always photograph—as an extension of my everyday life.

You capture beautiful moments with friends, family, and in your home. Your photos convey very personal stories.

H   I want to capture in photographs the things I encounter in life and the aspects of living that I find comforting. What I want to convey through my photos isn't a strong message but rather what I find beautiful and who I am as a person. The closer the subject is to my experiences and relationships, the more unique and personal my expression becomes.

服部恭平　写真家　モデルとしてパリや世界各国のランウェイショーやファッション誌で活躍。2018年よりライフワークでもあった写真を本格的に始動。2020年に写真集『2019-2020』を刊行。2021年には写真展「Anytime!」を開催。2023年には、新作のzineの刊行を記念した写真展「hug and photo」を開催した。

Kyohei Hattori worked as a model for runway shows and fashion magazines in Paris and around the world. In 2018, Kyohei pursued photography, which was a long-time passion, and in 2020, published his photo book *2019-2020*. In 2021, Kyohei held his photography exhibition *Anytime!* and in 2023, *hug and photo*, to commemorate the release of his new zine.

MIYU OTANI
Embracing Deep Emotions

Photography: Kasumi Osada
Hair & Make-up: kika
Text: Miyu Otani

agnès b. stories

分厚い想いを乗せて
小谷実由

Essay

agnès b. stories

agnès b. stories

撮影で偶然出会ったジーンズ。ずっとデニムパンツは大好きで、履いた時に自分が心地良いと感じたものをいつも選んでいる。でも、そのジーンズは当時やっとの思いで履けるくらい窮屈で、歩く時も膝があまり曲がらなくてぎこちなく、ご飯をお腹いっぱいに食べたら苦しくなってボタンを外したくなるほどストイックな履き心地だった。それでも惹かれてしまったのは、履いた時のシルエットに一目惚れしてしまったから。いつもよりもシャンと背筋が伸びて、気持ちまで真っ直ぐになれそう。いつかこのジーンズを心地良く履きこなし、まるで自分の身体の一部のようになれたとき、きっと今よりも自分のことが好きになれる気がした。早く"自分のもの"にしたくて、毎日のように共に日々を過ごした。それから3年、今ではぎこちなさも忘れ、スキップも軽やかにできそうな感じ。初めて履いた時に感じた気持ちは確信に変わり、このジーンズと一心同体になり始めている自分のことも好きだなぁと思える。この先もずっと一緒にいてほしい、なくてはならない相棒みたいな存在。

おはよう。ご飯だよ。いってくるね。ただいま。おやすみ。と、いつも声を掛けている愛猫。もうすぐ一緒に暮らして8年。もちろん血も繋がっていないし、何を考えているのかもわからないけど、彼の生活を見続け、日々の成長や変化を感じていると、心なしか何か自分に通ずるものがあるような気がしてくる。猫と人間なんだから似てるわけがないのに、同じような顔をしているように見えたり。天気が悪い日はなんだかやる気が出ない……と思うと彼もぼんやりとして床でゴロゴロしている。普段はこちらには興味がないような素振りなのに、時折私がしくしく泣いていると近くに寄ってきたりして、もしかして何かわかってるのかなぁなんて都合よく解釈しちゃって。そんな風に考える瞬間が年月を経て増えるたびに、愛おしさも増していく。いつか、一度でもいいから私の声掛けに返事をしてくれたらいいなと密かに願っている。

実家を出てから長年住んでいた大好きな場所を、やむを得ず離れなくてはいけなくなった。離れることが決まっても、離れてからも寂しい気持ちは大きくなるばかり。その場所にいる人々も、朝から晩までいつも清々しかった街の景色も、どれも自分を前向きにさせてくれるものだったから。しかし、いつまでも落ち込んでいるわけにはいかない。新しい場所でも良いなと思えるものを探すことにした。新しい家で迎える朝の空気は前よりも静かでゆっくりしていて、一日の始まりが穏やかに。少しだけ広くなったリビングはどこに座っていても落ち着かなかったけど、ふと大の字で床に寝転がってみたら気持ちが良くてなんだかこれも悪くないなと思い始めた。時間があると近所をぐるぐると歩き回った。最初は道も覚えられず迷ってばかりだったけど、日を追うごとに、同じ道を何度も通るたびに、好きだなと思える景色が増えてきて、落ち着くなと頭の中でぽつりとつぶやく日が増えた。馴染むというのは、もしかしたらこういうことなのかもしれない。長年住んでいたあの場所が大切という気持ちは変わらないまま。今の場所はそれとはまた違った形の大切という気持ちを抱き始めている。

大切に思うものは、日々を共にしているとその想いが深まっていく。いつか私がおばあちゃんになったとき、この大切に思うものたちはどれほど分厚い想いを乗せた姿になっているんだろう。

I came across these jeans by chance during a photoshoot. I've always loved denim pants and choose ones that make me feel comfortable. However, these jeans were so tight I could barely fit into them, and they were so stiff that my knees hardly bent when I walked. If I ate too much, I felt I needed to unbutton them. Despite this, I was irresistibly drawn to them because I fell in love with their silhouette at first sight. They made me stand up straighter than usual, uplifting my mood. I felt that if I could eventually wear these jeans comfortably, as if they were a part of my body, I would probably like myself more than I do now. Eager to make them 'mine', I wore them almost every day. Three years later, I no longer remember the awkwardness, and I could even skip lightly in them. The feelings I had when I first wore them have turned into certainty, and I like the version of myself that is becoming one with these jeans. I want them to stay with me forever, like an irreplaceable partner.

Good morning. It's time for food. I'm off. I'm home. Goodnight. These are the things I always say to my dear cat. We have been living together for almost eight years now. Of course, we are not related by blood, and I have no idea what he's thinking, but watching his life, and observing his daily growth and changes, I somehow feel a connection between us. Although a cat and a human can't resemble each other, there are moments when we seem to have the same expressions. On days when the weather is terrible, and I feel unmotivated, he also seems to be lounging lazily on the floor. Even though he usually acts uninterested in me, there are times when he comes close as I'm quietly crying, and I can't help but wonder if he somehow understands. As these moments increase, my affection for him grows even stronger. I secretly hope that one day, even if just once, he will respond to my calls.

After moving out of my family home, I had to leave a place I had loved and lived in for many years. Even after deciding to leave and when I was actually leaving, the feeling of loneliness only grew stronger. The people in that place, the refreshing scenery from morning to night; everything there always motivated me. But I can't stay sad forever. I decided to look for things I could appreciate in my new location. The morning air in my new home is quieter and more relaxed, making the start of each day calm. The slightly larger living room initially felt unsettling no matter where I sat, but one day, I sprawled out on the floor and found it surprisingly comfortable. It wasn't so bad after all. Whenever I had time, I walked around the neighbourhood. At first, I couldn't remember the directions and got lost all the time, but as the days went by, and I walked the same paths repeatedly, I began to discover more and more scenes I liked. I felt at ease and found myself quietly murmuring in my head that this was nice. Maybe this is what it means to feel at home. My feelings of attachment to the place I lived in for many years haven't changed. But now, I'm starting to develop a new sense of attachment to my current place, in a different way.

The things we hold dear grow significant as we spend each day with them. I wonder how much deeper these feelings will become and how enriched these cherished items will be by the time I become a grandmother.

小谷実由　モデル　14歳からモデルとして活動を始める。自分の好きなものを発信することが誰かの日々の小さなきっかけになることを願いながら、エッセイの執筆、ブランドとのコラボレーションなどにも取り組む。猫と純喫茶が好き。通称・おみゆ。2022年7月に初の書籍『隙間時間』(ループ舎)を刊行。

Miyu Otani began her career as a model at the age of 14. Alongside modelling, Miyu also writes essays and collaborates with brands, hoping that sharing what she loves can become a small inspiration in someone's daily life. Miyu loves cats and traditional coffee shops. Her nickname is 'Omiyu'. She published her first book, *Sukima Jikan* (Loop-sha) in July 2022.

JUNICHI HIGASHI agnès b. Japan Sales Dept.

アニエスベーを支える営業統括本部の東 淳一。憧れのファッションアイコンだったというアニエスベーに入社して27年、現在は営業として、ブランドの魅力を伝えるための店舗サポート、取引先とのコミュニケーション等の業務を担当している。スタッフはもちろんのこと、多くのカスタマーにも愛される彼の、思い入れのある大切なアニエスベーのアイテムたち。心温まる思い出とともに紹介する。

Junichi Higashi, Head of Sales at agnès b., joined the company 27 years ago because of his admiration for the brand as a fashion icon. Now, as a sales executive, Junichi is responsible for supporting stores, communicating with business partners, and promoting the brand. Loved by agnès b. staff and numerous customers alike, Junichi has a collection of treasured agnès b. items, each with its own heartwarming story. Here, we introduce these beloved pieces along with the memories they hold dear.

Clothes that Bring Energy

エネルギーをもたらす服

東 淳一

入社20年目に、パリの本社を訪れた時の出来事。本社でもなかなか、お目にかかることのできないアニエスがふらっと現れたんです。これはその時に着用していたカーゴパンツ。アニエスが、自宅の庭で見つけた落ち葉をプリントしたオリジナルの生地で出来たものです。アニエスは、私のパンツを眺めて、「その落ち葉は、私の庭にあったものよ」と笑顔で語りかけてくれたんです。20年越しにパリで会うことができた、夢のような時間でした。

In my 20th year with the company, I visited the Paris headquarters. Agnès, who is rarely seen even at the headquarters, appeared unexpectedly. These are the cargo pants I was wearing at the time. They are made from original fabric printed with fallen leaves that Agnès found in her garden. Agnès looked at my pants and smiled, 'Those leaves are from my garden.' It was a dreamlike moment to finally meet her in Paris after 20 years.

実は夫婦ともに、アニエスベーの社員。2004年に結婚した際、サプライズで世界にふたつしかないTシャツをプレゼントしていただきました。アニエスからの祝福メッセージ入りで、人生の節目を彩る忘れられない思い出になりました。

Actually, both my partner and I work for agnès b. When we got married in 2004, we were surprised with a gift of two unique T-shirts, the only ones in the world. They came with a congratulatory message from Agnès herself, making it an unforgettable memory that marked a significant milestone in our lives.

セットアップやスーツは、数多くコレクションしています。アニエスはアフリカが大好きで、アフリカンバティックをセットアップにするセンスが素敵だなと思います。フューシャピンクのコーデュロイのセットアップは、2006年に発売されたもの。この年のコレクションテーマはパンクロックで、この他にも魅力的なアイテムがたくさんありました。何度も着ていて、生地の汚れがたくさんある為、着ることは難しくなってしまいましたが、ずっと大切にしているアイテムのひとつです。

I have a vast collection of suits. Agnès loves Africa, and I find her sense of using African batik for suits wonderful. The fuchsia pink corduroy suit was released in 2006. The collection theme that year was punk rock, with many other captivating items. I've worn it many times, and the fabric has many stains, making it difficult to wear now, but it remains one of my treasured pieces.

入社した1997年に購入した、牛革のピーコート。27年経った今でも、現役です。時間をかけて、自分の身体に馴染んでいく革の質感。着るたびに初心を思い出させてくれます。

The cowhide pea coat I purchased when I joined the company in 1997 is still going strong 27 years later. The leather's texture has gradually conformed to my body over time. Every time I wear it, it reminds me of my beginnings.

中古の16ミリカメラで身の回りの撮影をはじめ、数々の日記映画を残した、映像作家のジョナス・メカス。1960〜70年代は商業映画が中心で、個人が撮影する日記映画は珍しく、同時期に活躍していたポップアートの旗手、アンディ・ウォーホルにも大きな影響を与えました。そのウォーホルが写ったフィルムがプリントされた、カットソー。ハイネックになっているところがポイントで、ウォーホルがハイネックを好んで着ていたことから、デザインに落とし込まれています。インディペンデントな精神が詰まったアイテムです。

Using a second-hand 16mm camera to start filming his surroundings, filmmaker Jonas Mekas left behind numerous diary films. In the 60s and 70s, commercial films dominated, and it was rare for individuals to make diary films. Mekas greatly influenced Andy Warhol, the leading figure of pop art who was active during the same period. This cut-and-sew top features a printed film of Warhol. The high neck is a key feature, reflecting Warhol's preference for high-neck tops. This item is imbued with an independent spirit.

アニエスベーのアイテムは、200着程もっています。その中でも一番のお気に入りなのが、JonOneのセットアップとシャツ。新店舗がオープンする際や、イベント開催時に必ず着用する特別なアイテムで、ワクワクする気持ちと心地よい緊張感をもたらしてくれます。アニエスベーはモノトーンのイメージがありますが、アーティストとのコラボレーションやカラフルなアイテムもあることを知っていただけるきっかけになれば嬉しいと思っています。ただカラフルなだけでなく、JonOneの作品がプリントされているということ。アートに溢れたエネルギッシュな洋服にいつも力をもらっています。

I have around 200 items from agnès b., but my absolute favourite is the suit and shirt by JonOne. I wear them on special occasions, like a new store opening or an event. It brings me excitement and pleasant tension. While agnès b. is often associated with monochrome, I hope this helps people discover the colourful items and collaborations with artists as well. It's not just about being colourful; the fact that JonOne's artwork is printed on them means these clothes are filled with energetic art, always giving me strength.

フランス人のル・モジュールドゼールの作品がプリントされたスニーカー。ストリートアートを主として活動しているアーティストです。初めてメンズコレクションの買い付けに参加させていただいた時に発売されたもので、記念に購入しました。2023年のコレクションにも、彼の作品を用いたアイテムが登場しています。

These are sneakers printed with the work of French artist Le Module de Zeer, who mainly works in street art. I bought them as a souvenir when I participated in the men's collection buying trip for the first time. The 2023 collection also features items using his artwork.

東 淳一　アニエスベー営業統括本部　1997年入社。アニエスベーの営業に関わることから、ブランドの魅力を伝えるサポートまで、幅広い業務に従事。最近では、2023年4月にオープンした、アニエスベー祇園店の立ち上げにも携わる。カラフルなスタイリングがトレードマーク。

Junichi Higashi joined agnès b. in 1997. Junichi is involved in a wide range of work, from sales to supporting the promotion of the brand. Recently, Junichi was involved in the launch of the agnès b. Gion store, which opened in April 2023. His trademark is his colourful styling.

デニス・モリスのキャリアは偶然の出会いに溢れていた。その中でもミュージシャンのボブ・マーリーとの出会いは重要なものと言える。デニスは、ボブのツアーなどに同行し、レンズを通して記録し続けてきた。アニエス・トゥルブレもそんなボブの音楽に魅了されたひとり。1975年にオープンしたアニエスベーの最初の店舗では、彼の音楽が繰り返し流れていたという。アニエスベーが日本上陸40周年を果たした2023年、ボブが世界的に注目されるきっかけとなったアルバム『Catch A Fire』も発売から50周年を迎える。デニスの個展「Portraits of the King」もそんな記念すべき年に青山のアニエスベー ギャラリー ブティックで開催されることになった。個展のオープニングに合わせて来日したデニスに、展示中の作品とそれらにまつわる当時の思い出について話を聞いた。

The Magic of Coincidence

偶然の魔法
デニス・モリス

ボブ・マーリーは僕が写真家として成長する姿を見守ってくれたし、僕も彼がミュージシャンとして成長するのを見守ってきた。当時、ジャマイカとイギリスの西インド諸島からの移民コミュニティ以外では、レゲエ音楽はまだまだ人気がなかった。小さなバスに乗って、半分も埋まらない会場を巡ってパフォーマンスする生活だ。それから数年後、世界中でボブのショーはソールドアウト。僕が撮影した彼のポートレイトが国際的な雑誌の表紙を飾るまでにもなった。ボブは僕に軌跡を残すことが大切だといった。彼の音楽は彼の哲学を伝える手段であり、僕の写真は若い黒人アーティストの道しるべになるべきなのだと。

写真を撮りはじめた頃は、戦争写真家になりたいと思っていた。それが、ボブとの出会いですべてが変わったよ。僕はルポルタージュのスタイルを音楽写真に取り入れて、それを新たな芸術としてアプローチしたいと思ったんだ。僕にとって写真とは、音楽やパフォーマーそのものに向き合うためのツールだった。ミュージシャンとして全盛期のアーティストを撮影することで、多くのことを学んだ。ボブから学んだのは精神性、そして自分自身のアイデンティティについて深く知ることの大切さ。

ボブのポートレートを撮影するのは、画家が一枚の肖像画を描き上げるのに似ている。撮影時は彼と僕だけ。簡単な会話を交わしてお互いのムードをつかみ合う。僕はカメラを手に取り、シャッターを切ってはカメラを置く。それを何回か繰り返す。ボブにポージングをお願いしたことはないんだ。彼に向かって「もう一度同じ動きをしてくれる？」なんて頼めないよ。お互いに万全の状態で臨むんだ。僕が撮影したボブの写真はどれもスタジオやライトは使わず、自然光を利用している。なぜなら、ボブのように光を放っているアーティストに照明なんて必要ないから。

ボブは自由は与えられるものではなく、自分で勝ち取るものだと教えてくれた。演奏する音楽から着る服に至るまで、すべてが彼のメッセージを反映していた。彼は特にデニムを愛用していて、スニー

シャンは、迷彩やカーキを取り入れたミリタリールックに夢中で、流行も生み出した。ジャマイカの人たちは、いつもシンプルでとてもスタイリッシュ。ボブも特別な格好をする必要がなかったんだ。何を着ても似合っていたよ。

魔法は存在する。占いとか、タロットとか、そういう類いではなく、意味のある人との決定的な出会い、それこそが魔法だ。僕は人生とキャリアを通して、いつでも自分の直感を信じて、適切なタイミングでどこにいるべきかを感じ取ってきた。

パリに住んでいた頃、僕と妻はいつもアニエス・トゥルブレが最初にアニエスベー ギャラリーを構えたジュール通りを散歩していた。散歩中いつも、アニエスベーのギャラリーで展示できたらいいねと話していた。そして2023年の春には「KYOTOGRAPHIE 京都国際写真祭」でアニエスベーの協賛のもと「Colored Black」展を開催し、パリのアニエスベー ギャラリー ラ・ファブでも同展示を開催した。そして、青山のアニエスベー ギャラリー ブティックで新しい個展「Portraits of the King」を開催した。こんな巡り合わせってあるんだね……。

アニエス・トゥルブレにはじめて会ったのはパリの彼女のオフィスだった。彼女が最初に話してくれたのは、ボブの伝説的なライブがパリではじめて開催された時に行ったこと、そしてそれが彼女の人生を変えたということだった。当時、多くの人々がそうであったように、ボブは、アニエスにインスピレーションを与えたんだ。僕は、長年彼女の洋服が大好きだった。トカゲがプリントされた白いスニーカーを今でも持っているよ。お店で購入した時、それが最後の一足だと知って、履き潰さないように、大切にしているんだ。

今までさまざまなレジェンドと仕事をしてきたけど、アニエス・トゥルブレはボブと同様にその中でも特別なレジェンドのひとり。彼女はデザインするだけでなく、カルチャーも生み出している。審美眼もあって素晴らしい写真のコレクションも持っているんだ。僕は、写真を通して世界中を旅し、素晴らしい人たちに出会ってきた。アー

Dennis Morris' career was marked by chance encounters. Among them, his encounter with Bob Marley could be described as decisive—Dennis has accompanied Bob on tours and continued to document him through his lens. Agnès Troublé is also one of those who has been captivated by Bob's music. In her first agnès b. shop that opened in 1975, Bob Marley's records were played on repeat. 2023 marked the 40th anniversary of agnès b. in Japan, and also the 50th anniversary of *Catch A Fire*, the album that brought international attention to Bob Marley. Dennis Morris' last exhibition *Portraits of the King* at agnès b. galerie boutique in Tokyo was held coinciding with those anniversaries. On this occasion, we talked with Dennis, who came to Japan for the exhibition's opening, to share with us his memories from those times.

Bob Marley saw me grow as a photographer, in the same way that I saw him grow as a musician. From those days of his first tour in the UK, travelling in a small transit van, playing in half-empty venues, at a time when reggae music wasn't popular outside Jamaica and the West-Indies community in England, to the years of sold-out shows and worldwide success. From taking images of Bob on my own initiative, to having my images on the cover of international magazines. Bob told me that it was important to leave a trail behind you. His music was a vehicle to his philosophy, and I hope my images to be a trail for young Black artists.

My ambition was to be a war photographer. But when I met Bob, everything changed. I took my reportage influence into rock photography and approached it as an art form. For me, my photographs were studies: it was a tool to study the music itself and the people themselves. Through photographing musicians at pivotal times in music, I learned a lot about myself and the world. What I learned from Bob was spirituality, and knowledge of my own history.

My sessions with Bob were similar to sittings, like when a painter is doing a series of portraits. Whenever we did those sittings, it was just me and him. We would have a conversation. Bob would be checking me out, to see where I was, mentally, spiritually. While we were talking, if I saw something, I would pick up my camera, take my shot and put it down. I never actually asked him to pose. With Bob, I couldn't turn to him and say, 'Could you do that again?' He knew I was as ready as he was. None of my images of Bob were taken in a studio, and I always used natural light. There are particular artists that don't need lighting, because they bring the light themselves. Bob just brought the light.

Bob taught me that freedom isn't something that is given, freedom is something you take. Everything in him reflected his message, from the music he played, to the clothes he wore. He particularly loved denim, but also trainers, and sometimes a leather jacket. Reggae musicians also made camouflage and khaki quite trendy. They were very much into that military look. As Jamaicans, we have always been very stylistic in a very simple way: it's just the way we put things together. Bob never needed to dress in any particular way. Whatever he'd put on, he would look good in it.

Magic does exist. Not magic as in reading the cards or whatever, but decisive encounters with meaningful people. That is magic. Throughout my life and career, I always trusted my instinct and sensed where to be at the right time.

When we lived in Paris, my wife and I would always walk around rue du Jour, where Agnès had her first shop and her art gallery. For years, I would visit the gallery and the incredible exhibitions, and always thought it would be great to do a show with agnès b. It eventually became real– with the sponsor of agnès b., we held the exhibition *Colored Black* at KYOTOGRAPHIE in spring 2023, and also at La Fab. in Paris. A new solo exhibition *Portraits of the King* was also held at Aoyama gallery in Tokyo. I never thought such coincidences would happen...

The first time I met Agnès, it was at her office in Paris. And the first thing she told me was that she went to one of the first Bob Marley's legendary shows in Paris, and that it changed her life. It empowered her, just like it did for so many other people at that time. I have loved her clothes over the years. I have a pair of white trainers with a printed lizard. When I got them, it was the last pair available. So I'm very particular on how I wear them, because I don't want to wear them out.

I've worked with some giants in my career, and Agnès one of the super giants like Bob. Because it's not only about design with her: she's a cultural figure. And on top of that she has a great eye. She takes a lot of pictures, and some of them are printed on her collections. Through photography, I've travelled the world, and met some incredible people. Agnès is one of them. Agnès, what an amazing woman!

デニス・モリス　写真家　レンズを通して非凡な人物たちに深く迫る作品群を制作する英国人フォトグラファー。音楽シーンと密接な繋がりを持ち、これまでにボブ・マーリー、セックス・ピストルズ、マリアンヌ・フェイスフルのアルバム『ブロークン・イングリッシュ』のカバー写真など、象徴的で印象的な写真作品を制作している。

Dennis Morris is a British photographer who creates a body of work that looks deeply into extraordinary people through his lens. With close ties to the music scene, Dennis has created iconic and striking photographic works, including cover shots of Bob Marley, the Sex Pistols and Marianne Faithfull's album cover *Broken English*.

「こんにちは！」爽やかでフレッシュな挨拶が、その場を一瞬にして明るく心地よい雰囲気で包み込む。
そんなパワーが一ノ瀬メイにはある。差別やカテゴリーを無くし、
個々が心地よく生きられる社会を目指して、水泳と言葉を通して発信してきた彼女は、
モデルや俳優業など、その表現の幅を広げている。アスリート生活は、常に自分との対話。
自身の心と向き合ってきた彼女の言葉は、温かく、優しいエネルギーを与えてくれた。
引退して3年目を迎える彼女の、これまでと今、そしてこれからの話。

形を変えて、伝えつづけたい

一ノ瀬メイ（モデル, パブリックスピーカー）

Continuous Communication
through Different Forms
MEI ICHINOSE (Model, Public Speaker)

Photography: Yuka Uesawa
Styling: hao
Hair & Make-up: Yuka Toyama
Text: Megumi Koyama

アスリート生活を終えて最近はどのような生活を送っていますか？

I　今までは企業に所属し、さらにスポンサーがいる状態だったので水泳が仕事という感覚。引退後のアスリートは、そのまま所属企業に残って正社員として働く道を選ぶ人も多いのですが、私はそれを全部やめてフリーランスに。自分ができる活動をやってみたいと思い、心機一転スタートさせました。最近は、学校や企業などへ向けた講演会をしたり、東京に出向いてメディアに出演したりしています。

個人で何かを発信しつづけたいという気持ちが強かったのですか？

I　水泳をしていた時も、タイムを更新してメダルを取ることが目的ではなくて、自分の思いを伝える手段として水泳をやってきました。水泳を通して達成できたこともたくさんあって、達成できたからこそ、社会から差別やカテ

ゴリーを無くしたいという目的に向かって形を変えてやってみたかったんです。個人でどこまで一ノ瀬メイとしてやっていけるか、わたし自身が見たいというのがすごくありました。

言葉で発信するということを高校生の頃からされていましたね。きっかけはなんだったのでしょうか？

I　全生徒が参加する学校内のスピーチコンテストでした。そこで優勝したことを機に「全国を狙ってみない？」と先生から話があったんです。水泳を通して伝えたいことを発信したいと思っていたけれど、言葉を使わないと届けられないこともあるということを痛感していたときでした。泳ぐことで、誰かのハートに届けられることももちろんあるけど、言葉を使って話すことで届けられることも、もっとあるんじゃないかなって。

社会の反応に対してどのような葛藤がありましたか？

I　パラリンピック出場を目指し始めた時、同級生や周りは誰もパラリンピックの存在を知らなかった。そうなったら、何か成績を残さないと話を聞いてもらえないなと当時の私は思ったんです。あとは、人と違う体で生まれて生活していると、嫌なことを言われることもすごく多かったし、スイミングスクールに入れてもらえなかったり、そういうことをたくさん経験しました。こんなにもみんなと一緒なのに、見た目や思い込みで区別されるというのが、悔しかった。

発信しつづけることで、周りや社会の反応に手応えを感じられた瞬間もありましたか？

I　スピーチコンテストで全国優勝してから取材が増えました。水泳の成績や結果だけではなくて、この人は何か伝えたいことがある人なんだということを、メディアの人に知ってもらうことができたと思っています。自分は伝えたいことがあって、その手段として水泳をやってるんだということを、スピーチコンテストという違うツールで伝えられたことは、大きなターニングポイントでした。"水泳"と"言葉"の掛け算は自分にとって立体的に思いを伝えることができる表現方法だと思っています。

2023年3月に行われたTEDxKyotoと11月に行われたTEDxP&G Singaporeにも登壇されていました。最初の参加から半年のブレイクがありましたが、どのような心境の変化があったのでしょうか？

I　引退してしばらくは、元日本代表や過去の栄光にすがるのがすごく嫌だったんです。だから早く次を見つけないと、という焦りがどこかにあって。でも2回のTEDxを通して、自分と向き合っていく中で、水泳でやってきたことをもっと誇りに思っていい、一生懸命取り組んで掴んだ結果を手放そうとする必要なんてないと気づくことができました。水泳選手だった自分も、今の自分を作ってくれているし、誇っていいこと。その後に築き上げているものも素晴らしいし、過去と現在進行形の自分がひとつになった感覚がありました。

何事においても、ポジティブに循環させていく力を感じます。

I　今の社会って生産性や能力主義みたいなところがあるし、何かができるから価値があるとか、人より何かが優れているから存在意義があるとか。そういう風に考えてしまうような流れがあると思っていて。でもそうやって考えている限り、片腕が短い分、人より何かできなかったら、人より価値が低いと考えてしまうじゃないですか。でも、みんなが自分に対して、何かができる／できない以前に、存在しているだけで十分じゃない？みたいに自分のことを許せていたら、目の前の人も同じように許せると思うんです。人に対して何か感じたときは、内省のタイミング。イラっとしたり、こういうところはあまり良くないなと思うと、私ってそこにこだわりがあるんだなと気づく。心のリフレクションだと思ってるし、自分の中で、消化したりとか内省する種になるっていうのは常に思ってるかもしれないです。

ライフスタイルに、ヴィーガニズムを取り入れるようになったそうですね。発見や気づきはありましたか？

I　発信者として大事にしていることは、自分の思ってることと、発言していること、やっていることを矛盾なく一本の線の上に乗せるということです。きちんと体現できているかって、その言葉のパワーを左右すると思っています。私は、みんなが心地よく生きていける社会を作りたいとか、差別がない社会にしたい、自分だけじゃなく

て周りの人に対しても境目なく大事にできるようになりたいというのがずっとあって。自分なりに100％体現してきたつもりでした。だけど、ヴィーガニズムというものに出会った時に、こんなふうに他の命に優しくなれるんだと思いました。自分のビジョンを体現する幅が広がったし、自分の中に知らず知らず存在していた矛盾が解消されたことが1番の気づきでした。

衣装を決める際に「これはハッピー？」とおっしゃっていたのがとても印象的で、その問いかけこそが一ノ瀬さんらしいと感じました。

I　アスリート時代は、自分の目標や夢というのがはっきりあって、常にそれが物差しだった。何かを選択するというときに、それは私を表彰台に連れていってくれるのか、メダルをくれるのかどうかで生きてきたんです。引退後は本当に、真っ白でしたね。でもその時に、自分の大事にしたいものや価値観をわかっているとそれが物差しになるんだと気づいて。地球に優しくできることや、他の命をできる限り傷つけず優しくできるとか、それが今の自分の中にある物差しですね。

これからやってみたいことについて教えてください。

I　カテゴリーとか肩書きではなくて、"一ノ瀬メイ"で何がやれるかみたいなところに挑戦していきたい。あとは、言葉を使うと境目が生まれてしまうということが難しいと思っています。障害という言葉を使った途端に、障害がある／ないという境目が生まれるし、ヴィーガンって言葉を使った途端、ヴィーガン／ノンヴィーガンという境目が生まれてしまう。だから言葉で伝えると同時に、ビジュアルや言葉じゃない形で伝える。この両輪が私にとってはバランスがよくて。モデルのお仕事もすごく楽しいし、去年は演技も初めて挑戦させてもらいました。いろんな形で表現したり、前に出ていけることがすごく楽しくて、その幅をどんどん広げていきたいなって思ってるし、それが他の人の希望になったりしたら嬉しいなと思っています。

agnès b. stories　interview

'Hello!' With her refreshing and vibrant greeting, Mei Ichinose instantly brightens the atmosphere, enveloping everyone in a pleasant and comfortable mood. Through swimming and words, Mei has been advocating for a society where everyone can live comfortably, free from discrimination and categories. Expanding her scope of expression, Mei has also ventured into modelling and acting. As she enters her third year of retirement, Mei shares her past, present, and future with us.

How has your life been like since retiring from your athletic career?

I   Previously, I was affiliated with a company and had sponsors, so swimming felt like a job. Many athletes choose to stay with their company as full-time employees after retiring, but I decided to quit everything and go freelance. I wanted to explore activities I could do and start afresh. Recently, I have been giving lectures at schools and companies and travelling to Tokyo for media appearances.

Did you have a strong desire to continue expressing yourself individually?

I   Even when swimming, my goal wasn't just to break records or win medals but to convey my thoughts and messages through swimming. I achieved many things through swimming, and because of those accomplishments, I wanted to make a difference in eliminating discrimination and categories in society by changing what I do. I was also curious to see how far I could go as Mei Ichinose.

You've been expressing yourself through words since high school. What triggered you to do so?

I   It was a school speech contest that all students had to participate in. After I won the competition, my teacher suggested, 'Why don't you try the nationals?' At that time, I was deeply aware that while I wanted to convey my message through swimming, there are things that cannot be communicated without words. While swimming could touch people's hearts, speaking could convey even more.

What kind of social response did you struggle with?

I   When I first started aiming to compete in the Paralympics, none of my classmates or the people around me knew what the Paralympics were. I felt no one would listen to me unless I achieved something significant. Additionally, living with a different body from other people meant that I often had to deal with hurtful comments. It was frustrating to be treated differently just because of appearances or preconceptions despite being fundamentally the same as everyone else.

Did you ever feel a positive response from those around you or society by continuing to express yourself?

I   After winning the national championship in the speech contest, I noticed an increase in interviews and media coverage. It helped people in the media to see me not just as an athlete with swimming achievements, but as someone with a message to convey. It was a significant turning point for me to be able to convey, through a different tool like a speech contest, that I swim because I have something I want to communicate. The combination of 'swimming' and 'words' has become a multi-dimensional way for me to effectively convey my thoughts and feelings.

You spoke at TEDxKyoto in March 2023 and at TEDxP&G Singapore in November. There was a six-month break between these two talks. How did your mindset change during this period?

I   For a while after retiring, I disliked clinging to my past glory as a former national team member. I felt an urgency to find my next path quickly. However, through the two TEDx talks, I reflected on myself and realised that I should take more pride in what I achieved through swimming. I learned there was no need to let go of the results I worked so hard to attain. The part of me that was a swimmer until retirement has shaped who I am today and is something I should be proud of. What I am building now is also wonderful, and I feel a sense of unity between my past and present self.

You have a strong ability to create a positive cycle in everything you do.

I   In today's society, there's a strong emphasis on productivity and meritocracy. There's a tendency to believe that our value lies in what we can do or how we excel compared to others. For example, it might make me think that my inherent value is lower than that of other people because I have a shorter arm. However, isn't it enough for everyone to simply exist in this world,

regardless of our abilities? If we can forgive ourselves this way, I believe we can similarly forgive the person in front of us. When I feel something about another person, it's a moment for introspection. If something irritates me or I think something is not good, it makes me realise that I have some attachment or concern about that aspect. I see it as a reflection of my mind. I constantly think about how these moments become seeds for my growth and introspection, allowing me to process and understand myself better.

I heard that you adopted veganism as part of your lifestyle. Were there any discoveries or realisations?

I   As someone who shares their thoughts with the world, I believe it's crucial to align what I think, say, and do in a consistent line without contradictions. The power of words depends on whether you can genuinely embody them. I've always wanted to create a society where everyone can live comfortably, free from discrimination, and value others as much as themselves. I have always intended to embody this vision 100% in my own way. However, when I encountered veganism, I realised how it could allow us to be kind to other lives, too. It broadened my ability to embody my vision and helped me resolve the contradictions within myself that I hadn't even been aware of. This was the most significant realisation for me.

When deciding on your outfit for today, your question, 'Do I look happy in those clothes?' left a strong impression.

I   As an athlete, I had clear goals and dreams, which were always my benchmarks. Whenever I had to make a decision, I would ask if it would help me reach the podium or win a medal. After retiring, I felt like I was starting with a blank slate. During that time, I realised that if I understood what was important to me and my values, those could serve as my new benchmarks. Being kind to the planet and minimising harm to other lives are the standards I now use to guide my decisions.

What would you like to do in the future?

I   I want to challenge what I can do as 'Mei Ichinose' without any category or title. Another challenge is that words create boundaries. The word 'disability' establishes a division between those who have disabilities and those who don't. Similarly, the term 'vegan' creates a boundary between vegans and non-vegans. Therefore, while conveying my message through words, I also want to communicate through visuals and non-verbal means. This dual approach balances things for me. I really enjoy modelling work, and last year, I had my first experience with acting. Expressing myself in various ways and stepping into the spotlight is incredibly enjoyable, and I want to keep expanding. I would be very happy if my journey could be a source of hope for others.

agnès b. stories Interview

一ノ瀬メイ　モデル, パブリックスピーカー　1歳半から京都市障害者スポーツセンターで水泳を始める。2010年、当時史上最年少の13歳でアジアパラ競技大会に出場し、50m自由形で銀メダル獲得。2016年のリオパラリンピックでは8種目に出場し、2020年には200m個人メドレー（S9クラス）で世界ランキング1位。引退後はモデルや俳優など、表現の幅を広げ活動している。

Mei Ichinose began swimming at the Kyoto City Sports Centre for the Disabled at the age of one and a half. In 2010, at the age of 13, Mei competed in the Asian Paralympic Games as the youngest participant, winning a silver medal in the 50m freestyle. At the Rio 2016 Paralympics, Mei competed in eight events and in 2020, she was ranked world number one in the 200m individual medley (S9 class). Since retiring, Mei expanded her career into modelling and acting, broadening her range of expression.

「美しく、大切な物語」。ニース、そしてパリのブティックで販売アシスタントとして勤めた後、現在はパリオフィスでプレスとして勤務するジャン＝ギョーム・ロベールは、17年のアニエスベー人生をそう表現する。パリのルデューにあるアニエスベー本社オフィスを訪れ、ブランドにまつわるさまざまな思い出深いストーリーを聞いた。

'A beautiful and long story.' That's how Jean-Guillaume Robert describes the 17 years spent at agnès b., from working as a sales assistant in the boutiques of Nice and Paris, to working as a press agent at the office in Paris. From the press room on rue Dieu, Jean-Guillaume shares some memories of a story that is not only beautiful and long-lasting, but also very personal.

b. yourself!

自分らしさを見つけた場所
ジャン＝ギョーム・ロベール

ラッパーのヤシーン・ベイ（旧名 モス・デフ）とアニエス・トゥルブレの出会いをセッティングする機会もありました。ふたりはすぐに意気投合して、アニエスはヤシーンに彼女がコレクションしている写真や私物などを見せていたのを覚えています。残念ながら、コラボレーションは実現しませんでしたが、今でも思い出に残っています。

I also had the opportunity to organise a meeting between the rapper Yasiin Bey (formerly known as Mos Def) and Agnès. I remember them getting on well together and Agnès showing Yasiin photographs and personal items. Unfortunately, this collaboration never came to fruition.

Photo: Delphine Migueres

アニエスベーでは日常的にさまざまなジャンルのアーティストと仕事をする機会に恵まれています。中でもダンサーのレオ・ウォークは印象的でした。2015年に本社オフィスのロビーで行われた彼のパフォーマンスに合わせて衣装を提供したんです。その後、アニエスベーの洋服をとても気に入ってくれて、テレビのパフォーマンスでも着用してくれました。スーツがとても似合っていたんですよ。

On a day-to-day basis, we're lucky enough to work alongside and dress many artists we appreciate. Among them, Léo Walk, whom we dressed in 2015 for one of his performances in the lobby of our headquarters at 15 rue Dieu. Léo loved agnès b. clothing, and he also borrowed some pieces for a performance on a TV set. The suit fitted him like a glove!

アニエスベーはカンヌ監督週間の長年のパートナーであり、カンヌ国際映画祭に出席する多くの俳優や監督に衣装を提供してきました。2022年に、幸運にも映画祭に参加する機会がありました。現在はアートギャラリーになっているラ・マルメゾンという18世紀に建てられた邸宅でアニエスベーのポップアップが開催されたんです。さらには、ビーチパーティも企画され、アニエス・トゥルブレ本人も出席しました。彼女は常に上質な映画を支援してきました。僕は、監督週間のオープニングを飾ったセドリック・カーンの『ゴールドマン裁判』とピエール・クレトンの『王子様』の上映にも参加しました。

agnès b. is a long-standing partner of the Directors' Fortnight and has always dressed many of the actors and directors present at the Cannes Film Festival. In 2022, I had the opportunity to go as a press attaché. An agnès b. pop-up was hosted at La Malmaison, an 18th-century mansion now serving as an art gallery. Additionally, we organised a beach party, which Agnès attended herself. Agnès has always supported auteur cinema. I had the pleasure of attending two official screenings: Cédric Khan's *The Goldman Case*, which opened the Directors' Fortnight, and Pierre Creton's *A Prince*.

Photography: Jean-Guillaume Robert, Text: Victor Leclercq

アニエスベーの服はどのくらい持っているだろう……。もう数え切れないほど。クローゼットがぱんぱんになるくらい！ サイズが合わなくなったアイテムは手放すけれど、逆に父親が着れなくなったアニエスベーのシャツを譲ってくれるので、親子でアニエスベーを愛用しています。家族の思い出も詰まっているんです。

16歳くらいの頃、ニースのロンシャン通りではじめてアニエスベーに出会いました。そして、すぐに虜になってしまった。BGM、壁に飾ってある映画のポスター、家具、店員さんの親しみやすさ。すべてが相まってとても温かい空間に感じられたんです。自分にぴったりだと思うブランドに出会い、その2年後、18歳の夏休みにそのお店で働きはじめたんです。

I've lost count of the number of agnès b. items I own. wardrobe is full of them! The only reason I can't wear some them now is that I don't fit into them anymore. But my fat gives me the shirts that he himself can't fit into anymore. It' family affair!

When I was around 16, I discovered agnès b. on Longchamp in Nice. I immediately fell in love with the wo of agnès b. which felt familiar: the playlists, the film post hanging on the walls, the furniture, the friendliness of people in charge—everything combined to make it a wa place where I felt good. I had finally found the brand t suited me, and I started working in this shop at 18 dur summer vacation.

販売員であれ、プレスであれ、自分の仕事は他のブランドではできないと感じています。アニエスベーと自を結びつけているのはファッションではなく、世界観です。映画、音楽、アート、すべてに対する情熱を共有することができるのです。アニエス・トゥルブレは、さまざまなアートをサポートすることに全力を注いでいます。彼女のようなボスは他にはいないでしょう。アニエスベーの精神を一言で表すなら「b.yourself!(自分らしく)」だね。

For a long time, I thought I wouldn't be able to do thi job, whether as a sales assistant or press agent, fo another fashion house. It's not fashion that connects me to Agnès, but rather the world we share. We share a passion for cinema, music and visual arts. What's more, Agnès is committed to serving others, one of the rare left-wing 'directors' of our time. If I had to sum up agnès b.'s state of mind in two words: *b. yourself*!

2023年でもっとも印象的だった出来事は、フランスのサイクルウェアブランド、カフェドシクリステとアニエスベーのコラボレーションのローンチイベントのために友人たちと20種類以上のピザラディエール（玉ねぎとアンチョビを使ったプロヴァンスの名物料理）を作ったことです！カフェドシクリステはニースに拠点があるので、ニース出身の自分がこの料理を作ることに必然性を感じました。40キロの玉ねぎの皮を剥き、オフィスの6階にある小さなキッチンで24時間料理しました……。幸運なことにたくさんの人に手伝ってもらい、ピザラディエールは大好評でした。手伝ってくれたみんなに感謝したい。これこそアニエスベーの精神だよ！

One of the most memorable events of 2023 was making over 20 *pissaladières* (a specialty from Nice made with onions and anchovies) with friends for the launch event of the collaboration between agnès b. and the French cycling apparel brand, Café du Cycliste. It felt natural to make this dish since Café du Cycliste is based in Nice and I am also from Nice. We peeled 40 kilos of onions and cooked for 24 hours in the small kitchen on the sixth floor of the office. Fortunately, many people helped, and the *pissaladières* were a big hit! I want to thank everyone who helped. This truly embodies the spirit of agnès b.!

ジャン=ギョーム・ロベール　アニエスベー パリ プレス　フランス・パリの本社にてプレスを担当。アニエスベーと出会ってすぐにファンになったジャン=ギョームは、そ年後にニースの店舗にて働くように。プレスとして映画や音楽、アートに関わる業務に携わる。

Jean-Guillaume Robert is a press agent at the headquarters in Paris. After their first encounter, Jean-Guillaume instantly became a fan of agnès b. and two years later, be working at the store in Nice. As a press agent, Jean-Guillaume engages in work related to film, music, and art.

A Place to Be Free

KOM_I

自由でいるための場所
コムアイ

Photography: Yuto Kudo
Hair & Make-up: TORI
Text: Rei Sakai

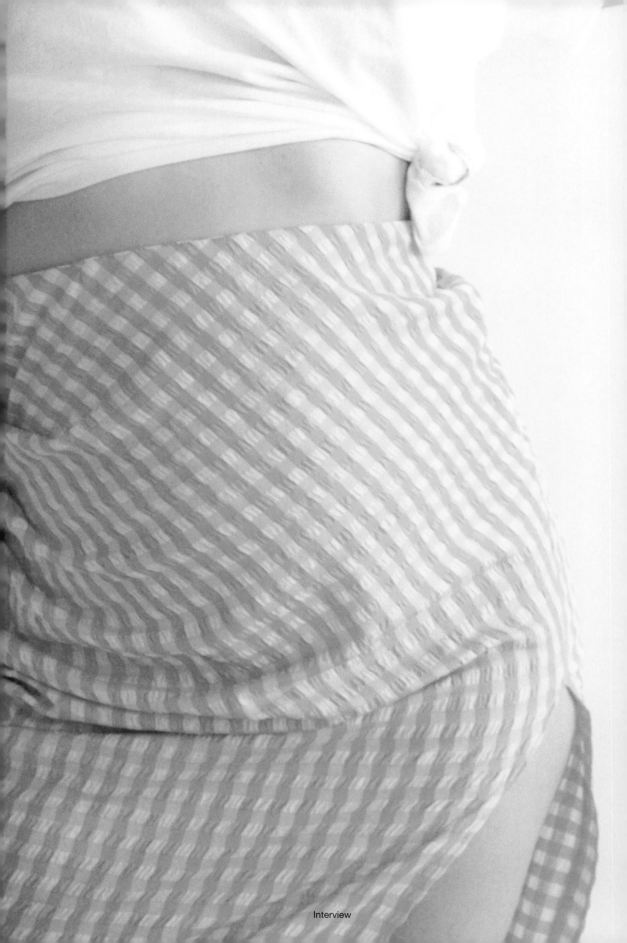

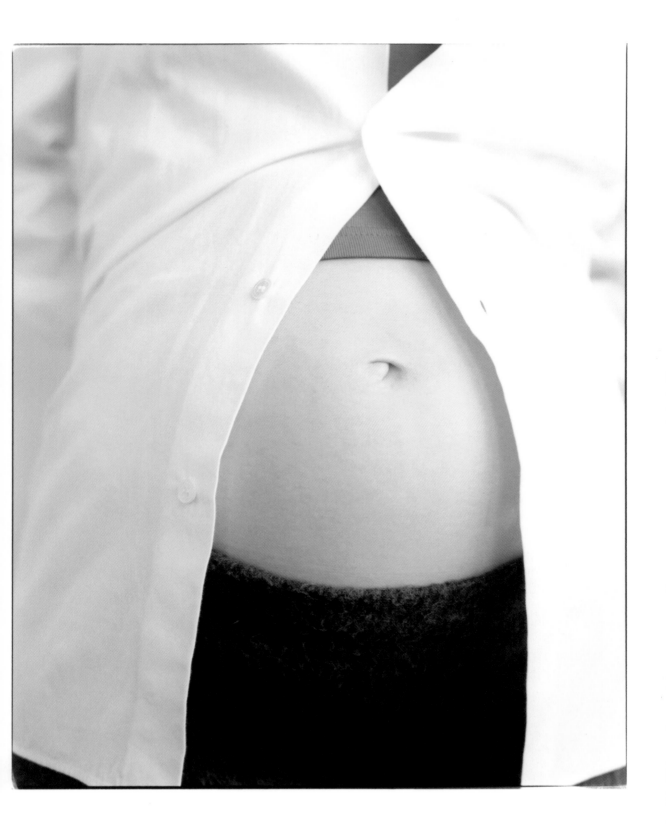

Interview

2023年、ソロ活動への転換や妊娠など、大きな変化の渦中にいたアーティストのコムアイ。
自身の表現を探求し続けるために、新たな挑戦を続ける彼女の芯にあるもの。その強さの秘訣を聞いた。

ソロ活動に転換してから今まで、変わらずに大切にしていることはありますか？

K　何事も即興だと思うこと。何百回も歌っている歌詞やメロディも、一年前から決まっているセリフも。すでに決まっているものをやるという意識でパフォーマンスすると、私は腑抜けになっちゃうんです。それにむずむずして違うことがしたくなってしまう。何でもいいんですけど、喜びや煌めきだけではなく、怒りでも、悲しみでも、憎しみでも。その時自分の心に現れた感情は決まっていたものではなく、新鮮なもの。だからそれを詰め込んで、声に出したり、踊りにしたり、表情に出すようにしています。外に出したら、結構味わえるものになることもある。パフォーマンスをするときは、何かしら感情の実が詰まっているような表現をするということを変わらずに大事にしています。

半定住のライフスタイルをしていると聞きました。インスピレーションには関係していますか？

K　それもあると思うんですけど、自分自身が生き物として動いていたいっていうのがある気がします。定住のように安定した場所を持つ方が落ち着く人がいるのと同じで、私にとっては不安定な方が合う感じがしている。ずっと同じ場所にいると約束する方が不安になるんです。

新しい暮らし方から得た気づきは？

K　どこでもホームにするということ。自分の身体と精神は常にあるわけじゃないですか。そこに、ホームを見出そうとしているのかなと思います。でもそれは、一人だけではやっぱり無理で。私は人が好きなので、一人で山の中に籠って暮らせるようなタイプではありません。そこに何か面白いコミュニティがあるから住んでみたいって思うタイプ。行った先とか、新しく進んだ先で、自分に合うものを諦めずに探して、居心地がいい場所を作る技術を鍛える。どこに行っても自分の好きな家を建てる方法がわかって、自分自身がホームみたいな感じにしたいなと思っています。

お子さんが産まれたら、今後の生活スタイルに変化はありそうでしょうか。

K　これはパートナーとも同じ意見なのですが、子供がいるからこれを制限しよう、ということになるべくならないようにと考えています。パートナーとの二人の関係が軸になっているのは変わらないので。楽しいから一緒にいるという気持ちを大切にしたくて、我慢してやってると思っちゃうと、仕事みたいになってしまうので。子育てについては、マナーとか社会的なプレッシャーの面での怖さもありますけど。

社会的なプレッシャーというと？

K　東京は、大人の社会に子どもが合わせるみたいな雰囲気が強いなと思います。電車に乗ってるときに静かに大人しくしてなきゃいけないとか。でもインドの電車で、子どもたちが遊んでて

動物園みたいな状態になっているのを見て（笑）。これだと子育てする人はもっと楽だなと思いました。知らない大人も子どもたちをすごく可愛がっていて。東京の空気みたいなものかな、何か申し訳ないっていう気持ちになることが多いのが怖くて、少しずつ変わったらいいなと思ってこういう場でも発信するようにしています。

パートナーである太田光海監督のドキュメンタリー映画『La Vie Cinématique 映画的人生』は、コムアイさんにとってどのような意味を持ちますか。

K　私、体験談は聞いたことがありますが、人の出産を実際に見たことがなくて。女の人も男の人も見ておくべきだと思うんですよね。そうじゃないと、自分の番になった時にイメージが湧かないし、リアルじゃないというか。ドキュメンタリー映画では何度か出産のシーンを見たことがありますが、なるほど、こうやって力むのかしら、と少し想像できたので、自分の番が来て妊娠から出産のプロセスを記録してもらい公開できることに、まず意義があると感じています。

新たな生命が自分の身体にあるってどんな感覚なのでしょう。

K　いまだに、人間の中で人間が育って出てくるって、すごく原始的だと思いませんか？人体でしか人体を育てられないんですよ。みんなへその緒がついていて、そこから栄養をもらっていたことがある。なんというか、結局のところ動物なんだなあ、って驚きます（笑）。人間ってこんなにプリミティブなものなんだという意識を持って世の中を見てみると、また現代が少し違って見えてくると思ってて。当たり前かもしれないけど、"作っている"感じがしないんです。だけど、胎児が自分で作っている感じもない。じゃあ何がそうさせてるのかというと、ただ作られているっていうことしかなくて、主体がわからない。受け身であるという感じなのかな。何か生命の流れみたいなものがあって、それに乗っかって自分たちが生かされてるだけっていうか。

全体の生命の流れに巻き込まれている感じ。

K　そう、巻き込まれてる！（笑）。結局はその一部という感じがします。遺伝子的に、パートナーの彼と私の遺伝子が入っているだけじゃなくて、ファミリーツリーを辿っていくと数億人の人間の遺伝子につながる。そして血のつながりのない個人の記憶も何かしら入っているのではと思えるんです。自分も産まれてくる子どもも、そういう色んな人の意識が入っているもので、でも一人の個人として生まれてくる。とても身近な他人と影響しあって生きていくのが楽しみです。

agnès b. stories

In 2023, KOM_I underwent significant changes: from transitioning to solo activities to experiencing pregnancy. We asked KOM_I about the core and secret of her strength as she continues to embrace new challenges in her pursuit of self-expression.

Since transitioning to solo activities, is there something you've consistently treasured?

K   Everything should be improvisational. Whether it's lyrics and melodies I've sung hundreds of times or lines that have been fixed for a year, I feel uninspired if I perform with the mindset that everything is already decided. It makes me restless and eager to do something different. It doesn't matter what emotion it is—joy, a sparkle, anger, sadness, or hatred. The feelings that emerge in my heart at that moment are not predetermined; they're fresh. So, I try to pack them in and express them through my voice, dance, or facial expressions. When I let them out, they often turn into something quite meaningful. When I perform, I always express something that feels like a cluster of emotions.

I heard you have a semi-settled lifestyle. Does it relate to your inspiration?

K   It does to some extent, but I also want to move around as a living creature. Some people feel more settled having a stable place like a permanent residence, but feeling unstable seems to suit me better. I get anxious when I commit to staying in the same place for too long.

What realisations have you gained from your new way of living?

K   It's about making anywhere feel like home. You know, our bodies and minds aren't always in the same place. We're always looking for some kind of home in that sense. But doing that alone is really difficult. I love being around people, so I'm not the type to live alone in the mountains. I'm the type who wants to live somewhere because there's an interesting community. Wherever I go or whatever new path I take, I want to keep searching for something that suits me without giving up, and I want to hone the skill of creating a comfortable place. I want to know how to build a home I like wherever I go and make myself feel at home.

Do you think your lifestyle will change when you have children?

K   My partner and I share the same opinion: we want to avoid restricting ourselves just because we have children. The core of our relationship remains the same. We want to cherish being together because it's enjoyable, and if it starts to feel like we're doing it out of obligation, it becomes like work. While there are concerns about manners and social pressures related to parenting, we hope to maintain our enjoyment and not let those factors dictate our lives.

What kind of social pressure are you thinking of?

K   There's a strong atmosphere in Tokyo where children have to adapt to adult society. Like, you have to be quiet and well-behaved on the train. But I saw children playing and the train turning into something like a zoo in India. (Laughs) Seeing that, I thought parenting would be much easier like this. Even other adults are very fond of children there. The unspoken air where people feel apologetic in Tokyo is scary, and I'm trying to gradually change that by speaking up in places like this.

What does the documentary film La Vie Cinématique by your partner, director Akimi Ota, mean to you?

K   I've never seen someone give birth myself. I've heard stories, but I think both women and men should see it. Otherwise, when it's your turn, you won't have an image in your mind, and it won't feel real. I've seen childbirth scenes several times in documentary films and thought, 'Ah, so this is how it happens', and could imagine it a bit. I thought that it would be meaningful to have the process of my pregnancy and childbirth recorded and shared publicly when it was my turn.

What does it feel like to have a new life in your body?

K   Isn't it incredibly primitive that humans are nurtured and born within humans? We can only grow humans within human bodies. Everyone has had an umbilical cord attached, receiving nourishment from it. It's like, ultimately, we're just animals, which is surprising! (Laughs) When you view the world with the awareness that humans are such primitive beings, it can make the modern world seem a bit different. It might sound obvious, but I don't feel like I'm 'making' anything. There's also no sense of the fetus making itself. So what's causing it? It's just being made, and the subject is unclear. It feels passive, like we're just being sustained by the flow of life and just riding on it.

It feels like we are caught up in the flow of life.

K   Absolutely, totally caught up in it! (Laughs) Ultimately, it feels like being a part of it. My grandparents passed away a few years ago. When my partner lived in the Amazon, the person who guided him like a father figure also passed away. When important people in your life are constantly being taken to the other world, you feel like you need to bring someone along somehow. Genetically, it's not just my partner's and my genes that are involved, but if you trace back our family trees, they connect to the genes of millions of other humans. And perhaps there are also memories of individuals with whom we don't share blood ties. Both myself and the child to come are made up of the consciousness of various people, yet we're born as individuals. It's exciting to think about how we'll interact and influence each other with those who are so intimately distant.

KOM_I　アーティスト　日本の郷土芸能や北インドの古典音楽に影響を受けながら、主に声と身体を用いて表現活動を行う。水にまつわる課題を学び広告する部活動「HYPE FREE WATER」を立ち上げるなど、環境問題にも積極的に取り組む。また、2021年からソロ活動へ転換。ドラマや映画に出演するなど、俳優としての顔も持つ。

Influenced by traditional Japanese performing arts and classical music from North India, KOM_I primarily engages in expressive activities using her voice and body. KOM_I launched 'HYPE FREE WATER' to address and promote awareness about water-related issues, actively tackling environmental problems. Since 2021, KOM_I has transitioned into solo activities, also appearing in dramas and films, showcasing her versatility as an actor.

Rock 'n' roll in agnès b.

服に宿るロックンロール

カキハタマユ

私がアニエスベーをはじめて手にしたのは高校生の頃。母がボーダーTシャツとカーディガンプレッションをプレゼントしてくれた時のことだ。その頃から私はロックに夢中でレコードやポスター、雑誌の切り抜きを部屋中に飾っていた。ニューヨーク・ドールズ、ザ・クラッシュ、ブロンディ、クランプス、ザ・モダン・ラバーズ、ビキニ・キル……。大好きなものに囲まれた部屋でレコードを聴くのが毎日の楽しみだった。

当時の私のスタイルと言えば、バンドTシャツにブラックのダメージデニム、ホールブーツを履くのが定番で、アニエスベーもカーディガンプレッションの中にブラック・フラッグやジョニー・サンダースのTシャツを合わせたりして少しパンクにアレンジして着るのが好きだった。その他にも、写真家のデヴィッド・ゴドリスが撮影したパティ・スミスの写真がプリントされたトートバッグ、ギルバート&ジョージのTシャツなど、アニエスベーの洋服はいくつも持っているけれど、どれも長く大切にしているものばかり。そんなアニエスベーを私がこれからもずっと着ていきたいと思うようになった大きな理由のひとつが音楽との関係だ。

デヴィッド・ボウイ、キム・フォーリー、トム・ウェイツ、サーストン・ムーア、キム・ゴードン、バクスター・デューリー、ジョン・グラントなど、数え切れないほどのアーティストへの衣装提供を行ってきたアニエスベー。ジャズの世界でもチェット・ベイカーがドキュメンタリー映画『Let's Get Lost 1988』の中でアニエスベーを着用していて、歳を重ねた彼の大人な魅力が引き立つ着こなしがかっこよかったし、デヴィッド・ボウイは1997年にマディソン・スクウェア・ガーデンで行われた自身の50歳記念ライブの衣装として着用し、プライベートでもアニエスベーを愛用していたそうだ。さらに、ソニック・ユースはアニエスベーの為に『AGNÈS B MUSIQUE』という曲を作っていた。およそ18分にもおよぶ長尺の曲で、浮遊感とノイズの共存した実験音楽だ。何故か気持ちが落ち着くような無機質なアンビエントミュージックで、キャッチーな曲でないからこそ、ショップでこの曲が流れていたりしたらすごく粋。

それにしても、こんなにも自分の好きなアーティストや音楽とアニエスベーに多くの繋がりがあるなんて驚きだ。若手のミュージシャンや、インディペンデントのレーベルを積極的にサポートしていたり、オフィスに音楽担当者がいたり、ショップ内に試聴機を設置してあったりと、ファッションの枠に留まらない面白い取り組みを数々行っているアニエスベー。フランスに1号店がオープンした当時、お店ではラモーンズがかかっていたとか。アニエスベー青山店の中で横山 健さんのミニライブが催されている動画を観たことがあるけれど、それも印象深かった。はじまる前に、スタッフの方から「モッシュなどの危険行為禁止」のアナウンスがあり、それがまるで"フリ"だったかのようにライブがはじまった瞬間モッシュが起きる（笑）。天井が低くて、ダイブしてる人の靴が空中を歩いてるみたいに靴裏が天井についちゃって……。きっとそんな光景は他のファッションブランドではなかなかみられないと思う。自由でアヴァンギャルドな姿勢を忘れない、そんなブランドのスピリットにもすごく共感している。

アニエス・トゥルブレというデザイナー自身のことを調べるほど、彼女の音楽、そして芸術に対する情熱に驚かされた。彼女は自分が受け取ることのできる一番の贈り物は音楽であると語っていて、家はキッチンを含むすべての部屋に音響システムが組み込まれているとか。そんな音楽への愛情が深い彼女の手掛けたブランドだからこそ、今も昔も変わらず多くのミュージシャンやアーティストからも愛され続けているのだろう。音楽ファンとして、そしてアニエスベーのファンとして、これからもずっと音楽を愛し、音楽からも愛される。そんなブランドであり続けて欲しいと願っている。

first encountered agnès b. when I was in high school. My mother gave me a striped T-shirt and a snap cardigan as a present. At that time, I was obsessed with rock music and had my room covered with records, posters, and magazine cutouts. New York Dolls, The Clash, Blondie, The Cramps, Modern Lovers, Bikini Kill... Listening to records in a room surrounded by my favourite things was what I looked forward to every day.

Back then, my go-to style was a band T-shirt paired with black distressed denim and combat boots. I loved giving my agnès b. snap cardigan a punk twist by wearing it over a Black Flag or Johnny Thunders T-shirt. I also owned several other agnès b. items, like a tote bag featuring a Patti Smith photo by photographer David Godlis and a T-shirt by Gilbert & George. I treasured these pieces, and they have been well-kept over the years. One of the main reasons I've come to love agnès b. so much and want to continue wearing clothes from the brand is its deep connection with music.

agnès b. has provided clothing for countless artists, including David Bowie, Kim Fowley, Tom Waits, Thurston Moore, Kim Gordon, Baxter Dury, and John Grant, to name just a few. In the jazz world, Chet Baker wore agnès b. in the documentary film *Let's Get Lost* (1988), showcasing a stylish look that enhanced his mature charm. David Bowie famously wore agnès b. for his 50th birthday concert at Madison Square Garden in 1997, and he was known to wear the brand in his private life as well. Additionally, Sonic Youth composed a track titled *AGNÈS B MUSIQUE* specifically for agnès b. This approximately 18-minute-long piece is an experimental blend of ethereal sounds and noise, creating a soothing, inorganic ambient feel. It's not a catchy tune, but it would feel incredibly chic if you heard it playing in a shop.

It's astonishing how many connections agnès b. has with the artists and music I love. From actively supporting young musicians and independent labels to having a music coordinator in the office and installing listening stations in their shops, agnès b. engages in fascinating initiatives that go beyond the realm of fashion. When the first store opened in France, they played the Ramones in the shop. I remember watching a video of a mini live performance by Ken Yokoyama held at the agnès b. Aoyama store, which left a lasting impression on me. Before it began, a staff member announced 'Don't do any dangerous activities like moshing', which seemed almost like a prompt because as soon as the live performance started, the crowd began moshing. (Laughs) The ceiling was so low that when people crowd-surfed, their shoes touched it, making it look like they were walking on air. Such scenes are rare for other fashion brands. I deeply resonate with agnès b.'s spirit of maintaining a free and avant-garde attitude.

The more I learn about Agnès Troublé as a designer, the more I am amazed by her passion for music and art. She once said that the greatest gift she can receive is music, and she has installed sound systems in every room of her house, including the kitchen. This deep love for music has made her brand continuously beloved by musicians and artists alike, both past and present. As a music fan and a fan of agnès b., I hope the brand will continue to love music and be loved by it in return.

カキハタマユ　DJ, レコード店スタッフ　中学生の頃にレコードと出会い、60 〜 70年代の音楽を聴き始める。レコード店スタッフ、ミュージックセレクター、エディターとし 音楽に携わる日々を送る。ロックとレア・グルーヴが好き。

Mayu Kakihata discovered vinyl records when she was in junior high school and started listening to music from the 60s and 70s. Mayu spends her days immersed in music, workir

Music that Feels Alive

JIRO YANASE

Photography: Kodai Ikemitsu
Styling: Daichi Hatsuzawa
Hair: Mikio
Text: Rei Sakai

生きている音楽
柳瀬二郎

Interview

2023年に5thアルバム『馬』を発表した「betcover!!」の柳瀬二郎。彼が奏でる音楽やパフォーマンスには、再現不可能な混沌や緊張感が漂っている。その正体について話を聞けば、人間としての原風景に立ち戻ることが鍵であるとわかった。彼の目に映る原風景とは、具現化するために意識していること、今後の野望について。

「betcover!!」の楽曲を聴くと、生であること、それゆえの緊張感だったり、新鮮味が伝播してくる感覚があります。

Y　生感でいえば、ここ3作くらいはライブと同じやり方でレコーディングしています。普通はクリックとかメトロノームを聴いてひとつずつ録っていくんですけど、うちは一発録り。失敗できないし、ミステイクがあったとしてもそれを生かします。音がぶつかることもあるし、リズムが揺らぐこともあるんですけど、そういうものをすごく大事にしていて。なんでも、精査されてきてるじゃないですか。音楽しかり、映画、写真、街もそう。ライブ感で録るっていうことが、僕の中ではすごく動物的だと思っています。

綺麗なものだけが見える現代社会だからこそ、崩れそうな過程を愛している？

Y　そうですね。何よりも、人間の原風景を大事にしています。原風景があった上での人生なので。実家に帰ろう、みたいな感じですかね。人と会っても、その人の実家がどんなところかとかは分からないじゃないですか。僕が目指しているのは、いま現在の荒ぶった状況のことを描きながら、同時に原風景が見えるようなもの。そういう作品になっていたらいいなと思います。

現代社会で生活していると、日に日に原風景を失っていく感覚があるし、次第には自分が本能的に失いたくなかったものを判別しにくくなっているように思います。柳瀬さんにとって、失いたくないものがあるとすればそれは何ですか？

Y　いろいろあるんですけど、フランス映画は特にそうですね。ちょっとエロいやつ（笑）。中学3年の時、仮病を使って全然学校に行かなかったんですよ。実家で、1年間ほぼダラダラしてたんですけど、BSとかでよくフランス映画が流れてたんです。いま思うと巨匠の音楽家が曲を作っていて、そういうところで音楽を聴いてましたね。フランス映画って、何も起こらない。最終的に全部がどうでも良くなって全裸で花畑で踊るみたいな感じで、それこそ原風景というか、すごくいいなあって。大人が子どもに還るみたいな、そういう音楽とか感覚はすごく大事にしています。

いまの荒い状況と原風景を重ね合わせている感じでしょうか？

Y　そうですね。メロディは歌謡とか合唱曲を参考にしてたりするので、それにファズをかけて両立させたり。チャルメラとか、夕焼け小焼けとかって、みんなが幼少期に通ってるものだし原点じゃないですか。都会に行くとガシャガシャした音が鳴ってて、地方に行くと木とか動物がいて、その両方がブワーって混ざってく感じというか。楽器でガシャガシャして、合唱的な純粋なメロディーがあって、それをごちゃ混ぜにしてるみたいな。

言葉のインスピレーションはどこから来るのでしょうか。

Y　大体は情景描写なんですけど、想像でストーリーを作っています。状況だけ想像で作って、それ以外に自分の考えを反映する。小説とかも自分の経験していないことを書いてたりするわけで、音楽も別に自分の体験談を赤裸々に描く必要はないと思うんです。想像力とそこにワンポイントリアルがあれば、という感じです。

「betcover!!」のみなさんはスーツで揃えていますよね。緊張感で自分を包むというのを意識されていたりしますか？

Y　僕はその方が安心するんです。ラフな感じはあんまり好きじゃない。お客さんがいて楽しみに来てくれているので、現実離れしたスーツの男5人が楽器弾いてるの、いいなって。スーツだったらいくつになってもロックンロールできるじゃないですか。日本人でも海外の人でも、いくつになってもブチギレてる人が僕は好きなので（笑）。そういう感じをおっさんになってもやっていたいです。

Jiro Yanase, from betcover!!, released his 5th album, *Uma (Horse)* in 2023. His music and performances exude irreplicable chaos and tension. When asked about the essence of his music, it became clear that returning to the primal landscapes of human experience is key. We delve into what these primal landscapes mean to him, what he consciously focuses on to bring them to life, and his ambitions for the future.

When you listen to betcover!!'s music, you can feel the immediacy, tension, and freshness of the moment.

Y   Speaking of that rawness, we've been recording the same way we do live performances for the past three albums. Usually, you record each part while listening to a click track or metronome, but we do it in one take. There's no room for mistakes; even if there are, we use them. Sometimes, the sounds clash, and the rhythm wavers, but we value those imperfections. Everything is so scrutinised these days—music, movies, photos, even cities. Recording like a live concert is very primal and animalistic to me.

Is it because we only focus on beautiful things in our society today, that you find yourself loving the feeling to be on the verge of falling apart?

Y   Yes, exactly. Above all, I cherish the original landscapes of human life. Our lives are built upon these fundamental experiences. It's like returning to your childhood home. When you meet someone, you don't know what their childhood home is like. I aim to create works that depict the tumultuous nature of the present while also revealing those original landscapes. I hope my work can reflect that.

In modern society, it feels like we are losing our original landscapes day by day, and eventually, it becomes difficult to discern what we instinctively do not want to lose. If there is something that you didn't want to lose, what would it be for you?

Y   There are many things, but French films, especially the slightly erotic ones, are big for me. (Laughs) In my third year of junior high school, I pretended to be sick and hardly went to school. I spent almost a year lazing around at home, and during that time, French films were often aired on TV. Looking back, I realise that music by master musicians was being used in those films, and that's where I was exposed to a lot of music. In French films, nothing really happens. They often end with a scene where everything seems trivial, and the characters dance naked in a flower field. That's what I mean by primal landscapes—it's wonderful. It's like adults returning to childhood. The music and sensations I derived from those experiences are something I cherish deeply.

Are you overlapping the current chaotic situation in your music with those original landscapes?

Y   Yes, exactly. I often draw inspiration for melodies from traditional Japanese pop songs and choral music, then blend them with fuzz effects to merge both aspects. Tunes like 'Chārumera (Charamela)' or 'Yūyake Koyake (The Sunset Glow)' are about things everyone experiences in their childhood. In the city, you hear a lot of noise, while in the countryside, there are trees and animals, and I aim to capture the feeling of both merging. It's like combining chaotic sounds from instruments with pure, choral-like melodies and mixing them.

Where does your inspiration for lyrics come from?

Y   Most of it comes from descriptive imagery, but I create stories from my imagination. I imagine the situation and then reflect my thoughts on it. Just like novels often depict things that the author hasn't personally experienced, I don't think music needs to portray my own experiences vividly. It's about combining imagination with a touch of reality.

You all wear suits in the band, don't you? Is it a conscious effort to wrap yourself in a sense of tension?

Y   I find it more comfortable that way. I don't like casual, rough looks. Since our audience is coming to enjoy the show, I think it's nice to have five guys in suits playing instruments, which looks quite surreal. You can do rock'n'roll at any age with suits. I admire people, whether Japanese or from abroad, who are still passionately rocking out no matter how old they are. (Laughs) I want to keep that spirit alive even when I get older.

柳瀬二郎　ミュージシャン　ソロ・プロジェクト「betcover!!」のボーカルをつとめる。レーベル・事務所からの独立後、2021年にアルバム『時間』をリリース。2022年にはアルバム『卵』を提げ全国ツアーを開催。2023年10月ニューアルバム『馬』をリリースし、初の全国ワンマンライブツアーを開催した。

Jiro Yanase is the vocalist of the solo project betcover!!. After becoming independent from his previous label and agency, Jiro released the album *Time* in 2021. Jiro toured the country with the album *Tamago (Egg)* in 2022. In October 2023, Jiro released a new album *Uma (Horse)* and held his first nationwide one-man live tour.

RAN TONDABAYASHI Art Director

2023年に代官山T-SITE ガーデンギャラリーで開催された、アニエスベーがブランドの歴史を振り返るイベント、Signé agnès (b.)「アニエスベーを巡る」展で、ARを使ったアート作品を展示した、とんだ林蘭。このプロジェクトのためにパリへ渡った彼女に、アニエスベーを通したパリでの思い出や、今回のプロジェクトに対する想いを聞いた。

Ran Tondabayashi exhibited an AR-based artwork at the Signé agnès (b.) *Touring agnès b.* exhibition, an event held at the Daikanyama T-SITE Garden Gallery in 2023, in which agnès b. looked back on the brand's history. Ran travelled to Paris for this project. We asked her about her memories of Paris and her thoughts on this project.

Connecting with Agnès in Paris

アニエスゆかりの土地を訪れて
とんだ林蘭

パリの美味しい思い出
アニエスさんがよく通っていたという、カフェ・ド・フロールに行ってきました。コラージュムービーには、8つのパリの街並みを使用したのですが、その内の2つにカフェ・ド・フロールを登場させました。印象深い思い出のひとつが、フランス人のお友達とご飯を食べていた時の出来事。パリの人はメインをシェアする習慣がないらしく、1人ずつメインを食べようと言って選ばせてくれたんです。日本だと料理をシェアすることが多いじゃないですか。なのでそれがすごく新鮮で、これもいいなって思いました。

Delicious memories of Paris
We visited Café de Flore, a place Agnès frequented. In the collage movie, I used eight scenes of Paris, two of which feature Café de Flore. One memorable experience was dining with a French friend. Parisians don't usually share their main courses, so they suggested we each have our own main dish. In Japan, we often share our meals, so this was a refreshing change, and I thought, 'Oh, this is nice too.'

アニエスベーと学ぶ、パリ
レアール地区にあるアニエスベーの1号店など、アニエスベーにとって大切な場所を訪れながら、パリ中を散歩しました。アニエスベーと親和性のあるパリの街や空気を感じられたことが、今回の作品を制作をする上でとても大事なことでした。パリを拠点とするフォトグラファーのマルセロ・ゴメスさんに、作品に使用するパリの街並みを撮影していただきました。すごくチャーミングで優しさに溢れている方で、彼の視点を通して見るパリの街並みは新鮮でとても刺激的でした。

Learning with agnès b. in Paris
We visited agnès b.'s first store in Les Halles and walked around Paris, exploring places important to the brand. Experiencing the city and its atmosphere, which resonates with agnès b. was crucial for creating this piece. Paris-based photographer Marcelo Gomes captured the Parisian streets for the project. Marcelo is incredibly charming and kind, and seeing Paris through his perspective was refreshing and highly stimulating.

未来へ繋げるプロジェクトとして
展示ではARのブースとコラージュムービーのブースが別であって、同じモチーフを使ってそれぞれを構成しました。私が作っているムービーは、実はとてもアナログな手法を使っていて。モデルが動く姿を連続的に撮影して、1枚1枚写真を動かしながら作りました。ものすごく時間がかかることをやっているんですけど、その方が可愛くできる。一方でARは最新の技術なので、同じモチーフですごく対極というか、表現が違う楽しみ方ができたのかなと思います。

The unchanging spirit of Agnès
I was reminded that agnès b. is a brand that values a free spirit. Listening to Agnès, I realised that she isn't your typical designer in a good way; she does what she can freely and is full of creativity. Throughout this project, including the trip to Paris and the shoot in Tokyo, we were able to maintain our core values but were given a lot of creative freedom. I felt a deep respect for creators, and I think it's incredible that such a large brand allows for this level of creative liberty.

アニエスの変わらないスピリット
自由な精神を大切にされているブランドだということを、改めて感じました。アニエスさんの話を聞いていても、いい意味であまりデザイナーっぽくないというか、出来ることを自由にやっていらっしゃる、クリエイティビティに溢れた方なんだなと思いました。今回のパリ出張や東京での撮影も含めて、もちろん軸はブレないようにやっているけれど、クリエイティブ面はとても自由にやらせてくれて。クリエイターに対する敬意を感じましたし、こんなに大きなブランドでそれができるってすごいことだなと思っています。

A project connecting to the future
The exhibition featured separate booths for AR and collage movies, both using the same motifs. The movie I created employed analogue methods. I continuously photographed the model's movements and animated the photos frame by frame. The process was extremely time-consuming, but it resulted in a charming effect. On the other hand, AR utilises the latest technology, so it was interesting to see how the same motifs could offer two vastly different forms of expression.

とんだ林蘭　アートディレクター　コラージュ、イラスト、ペインティング、立体、映像など、幅広い手法を用いて作品を制作する。猟奇的でいて可愛らしく、刺激的な表現なスタイルとする。音楽アーティストや、ファッションブランドへも作品提供を行うなど、精力的に活動の場を広げている。

Ran Tondabayashi creates works using various techniques, including collage, illustration, painting, sculpture, and video. Her main style is characterised by expressions that are both grotesque and cute, as well as provocative. Ran actively expands her scope of activities by providing works for music artists and fashion brands.

RYO MATSUURA

How the World of Cinema Changed Me

Photography: Shota Kono
Styling: Tatsuya Yoshida
Hair & Make-up: Yoko Hirakawa
Text: Rei Sakai

〒604−8790

025

〈受取人〉
京都市中京区梅忠町9-1

株式会社 **青幻舎** 行

金受取人払郵便

中京局
承認

6239

（切手不要）

有効期間
5年6月15日まで

|||||||||||||||||||||||||||||||||||||||||||||||||||||||||||||||||||||||||||||

前（フリガナ）	性別	年齢
	男・女・回答しない	歳

所　〒

ail	ご職業

舎からの
・イベント情報を
しますか？
る　□しない

読者アンケートは、弊社HPでも
承っております。

最新情報・すべての刊行書籍は、
弊社HPでご覧いただけます。

青幻舎　　　　　検　索

https://www.seigensha.com

読者アンケート

お買い上げの書名	ご購入書店

本書をご購入いただいたきっかけをお聞かせください。

- □ 著者のファン　□ 店頭で見て
- □ 書評や紹介記事を見て（媒体名　　　　　　　　　　　　　）
- □ 広告を見て（媒体名　　　　　　　　　　　）
- □ 弊社からの案内を見て（HP・メルマガ・Twitter・Instagram・Facebook）
- □ その他（　　　　　　　　　　　　　　　）

本書についてのご感想、関心をお持ちのテーマや注目の作家、弊社へのご意見・ご要望
ございましたらお聞かせください。

お客様のご感想をHPや広告など本のPRに、匿名で活用させていただいてもよろしいでしょ

□はい　□い

ご協力ありがとうございま

アンケートにご協力いただいた方の中から毎月抽選で5名様に景品を差し上げます。当選
発表は景品の発送をもってかえさせていただきます。
詳細はこちら https://www.seigensha.com/cam

私を変えた映画の世界
松浦りょう

Actor

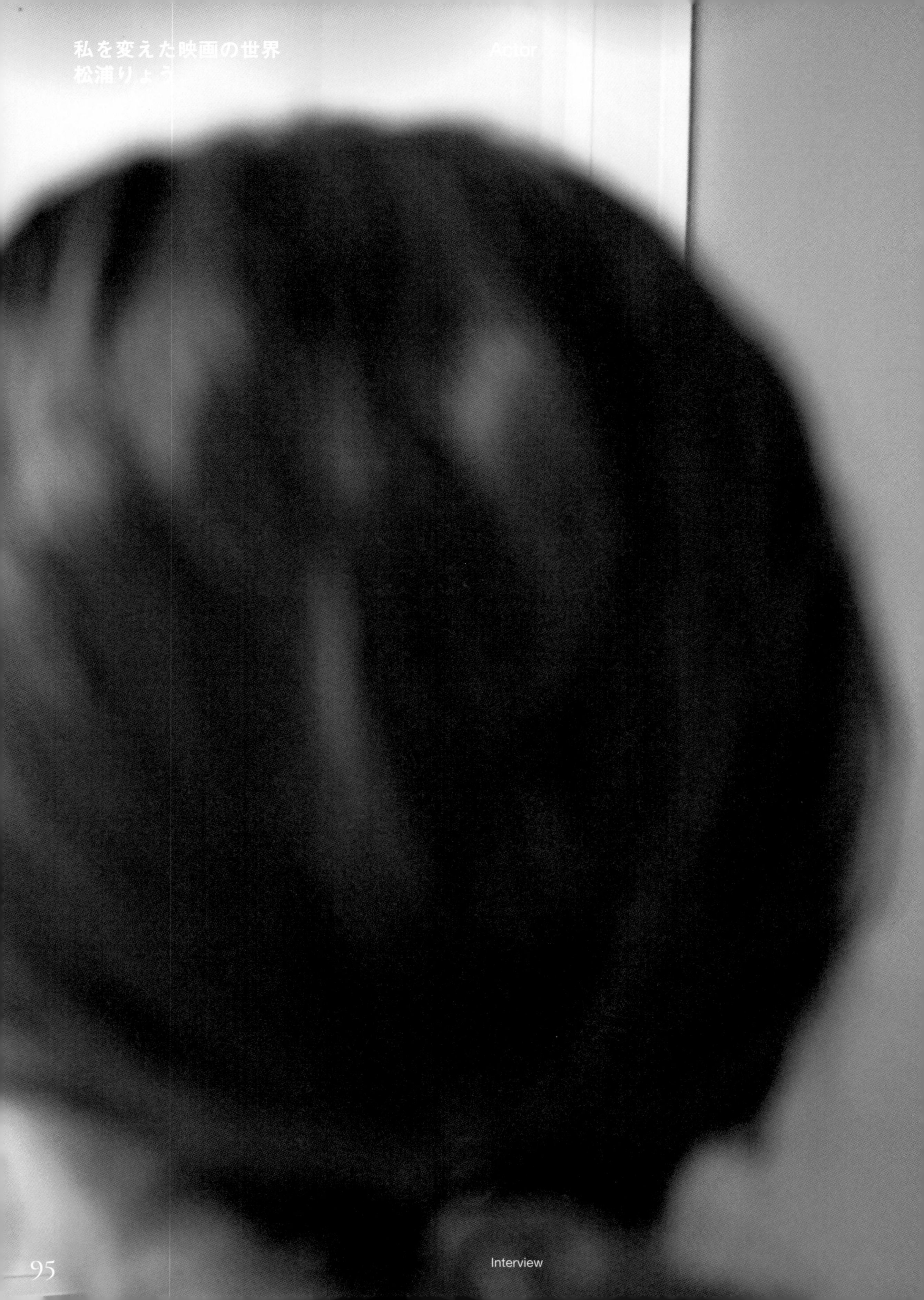

99

モデルから俳優へと活動の幅を広げ、2023年1月に独立を発表した松浦りょう。
演じることを通して向き合ってきた自身のコンプレックスや性質、それらを克服するための思考法や
人との向き合い方について。彼女がまるで取り憑かれたかのように役に入り込むその理由は、
自身が生きやすくなるためのひとつの手段だった。

今回は、アニエスベーのアイテムを取り入れた素の自分に近いスタイリングと、自分とは少し距離のあるスタイリングを着ていただきました。ジャケットにジャージのスカートは斬新な組み合わせですね。

M　どこでも寝転がれるような格好をするのが好きです。気持ち的に着ていて楽で、飾らない服。普段はボロボロのTシャツやジャージ、今日も小学生の頃に買ってもらったデニムを穿いて来たんですけど、古着の抜け感というか、緩いファッションが自分に合うなと思います。それは多分母親の影響で、小さい頃から古着を着せられていたみたいで。古着を着ていると自分のあるべき姿というか、元の姿になったように感じます。ジャケットはかっちりした印象ですが、今回選んだアニエスベーのジャケットは麻でできているので、ゆったりカジュアルに着こなせるのがいいですね。

もうひとつは、鮮やかな水色のノースリーブにタイトスカート（P97,P98）。シンプルですが一瞬で人を惹きつけるようなコーディネートです。

M　私、ウォン・カーウァイ監督の『恋する惑星』という作品を見たときに本当に一目惚れのように心を打たれて。いつかこの世界に入ってみたい！と思ったんです。このスタイリングはフェイ役を

演じるアーティスト、フェイ・ウォンの『チャン・ヨウ』という私の大好きなCDジャケットのビジュアルのイメージに近付けて組んでいただきました。私にとって彼女はずっと憧れの存在だったので、小さな夢が叶った感覚で本当に嬉しいです。自分とは少し距離があるという意味でいうと、タートルネックやタイトスカートのような、女性らしいアイテムはあまり持っていません。昔からメンズライクなスタイルが好きなので、サイズ感もゆったりしたものを選ぶことが多いです。でも最近はちょっとだけ女性らしさも入れたいなと思っていて、タイトスカートも今日のようなスタイリングだと可愛くて、素敵だなと思いました。

2014年から俳優を始められていますが、これまでに経験したことがない感情や、思い出したくないほどの気持ちを引き出さなくてはいけないこともあると思います。普段から、感情をコントロールすることは意識されていますか？

M　私はすごく感情豊かな人間で、私生活では感情のリミッターを外すようにしています。それがお芝居に対して生かされるからという理由ではなくて、ストレスを溜めることがすごく苦手なので、もちろん人を選びますが、自分の信頼している人に対しては、思ったことは基本的に何でも口にしていて。それこそ昔は、自

From modelling, Ryo Matsuura expanded her activities to acting and became independent in January 2023. Ryo talks about the complexities that she confronts through acting, how she thinks to overcome them and how she deals with people. Ryo gets into roles as if she was obsessed with them, which is one way to make her life more manageable.

This time, you are wearing a style that is close to your true self using agnès b. items, and another style that feels slightly different from your usual self. The combination of a jacket and a jersey skirt is quite innovative.

M　I like to dress in a way that feels comfortable anywhere. Comfortable and unpretentious clothing. Usually, I wear ragged T-shirts and sweatpants; today, I even came wearing the denim jeans I got in elementary school. Vintage, relaxed fashion suits me. It's probably due to my mother's influence, as she dressed me in second-hand clothes from a young age. Wearing vintage clothing feels like my true self, my original form. Jackets usually give a formal impression, but the agnès b. jacket I chose this time is made of linen, which makes it comfortable and casual to wear.

Another outfit was a vibrant light blue sleeveless top paired with a tight skirt (P97,P98). It's simple but has a captivating effect.

M　When I watched Wong Kar-Wai's film *Chungking Express*, I was struck as if I had fallen in love at first sight. I thought, 'I want to enter this world someday!' This styling was put together to resemble the visual from the CD jacket of *Chàng Yóu (Sing and Play)* — one of my favourite albums — by Faye Wong, the artist who played the role of Faye in the film. She's been an icon I've

always admired, so it feels like a small dream has come true. In terms of being a bit different from my usual self, I don't own many feminine items like turtlenecks or tight skirts. I've always liked menswear-inspired styles and often choose more relaxed fits. But recently, I've been wanting to incorporate a touch of femininity, and I found that tight skirts can be cute and lovely, like in today's styling.

You've been acting since 2014. There must be times when you have to evoke emotions you've never experienced or feelings you'd rather not remember. Do you consciously control your emotions on a daily basis?

M　I'm very emotional, and in my private life, I try not to control and limit my emotions. Not because it benefits my acting, but because I'm really bad at bottling up stress. Of course, I choose the people I confide in carefully, but with those I trust, I generally say whatever comes to mind. In the past, I used to express my feelings as they were, which led to a difficult rebellious phase and probably hurt a lot of people. Nowadays, I choose my words carefully and try to convey my feelings. I never completely suppress my emotions. The range of emotions I've accumulated over my life forms the basis of my acting. In that sense, I'm glad to be an emotionally expressive person.

分の思ったエネルギーのまま表に出していたので、反抗期もすごかったし、人をたくさん傷つけてきたと思います。でも、いまは言葉を選び、自分の気持ちをしっかり伝えるようにしています。自分の感情を絶対にゼロにはしない。これまでの人生で培ってきた感情の蓄積がお芝居の引き出しになっていると思うので、そういう点では感情豊かな人間でよかったなと思います。

エネルギーを出す難しさの反動で、感情を溜め込んでいた時期もありましたか？

M　ありました。究極の不器用なので、ストレスを過剰に感じた時は、全くコミュニケーションが取れなくなってしまうこともありました。自分のエネルギーをコントロールできないし、頭が真っ白になってしまい、場合によっては話すことさえできなくなる。俳優のお仕事は、私にとってとてもバランスが取れていると思います。お芝居って、役を通さなくてはいけないのですごく繊細な作業ではあるんですけど、お芝居を通して自分の培ってきた感情を表現できると、自己肯定感にも繋がるというか。私には合っているし、好きだなと思います。

2023年には独立されていますよね。何か大きなきっかけがあったのでしょうか。

M　私自身、事務所を辞めるなんて思ってもいなかったんですけど、2022年の10月に釜山国際映画祭に行ったんです。それまでは映画祭って楽しいお祭りのようなものだと思っていたんですけど、それだけじゃなくて。映画関係者しかいない貴重な集まりだからこそ、映画の宣伝はもちろん、そういう場でしか出会えな

い人との出会いもある。私は何の準備もせずに行って、英語も話せなくて。映画祭がどういう場所かということすら知らなかったんだって、後悔したんです。それが大きなきっかけで、いまのうちに自分でゼロからやることを経験するために独立しました。

実際フリーランスで半年過ごしてみて、どうですか？

M　難しいことだらけです。フリーランスだと信用問題でできないことがあったり、責任の重さとか、知らなかった事情がたくさんあって。いままでの何百倍も人に対して感謝できるようになりました。その一方で、直接お仕事をいただけることがこれまで以上に嬉しくて、ピュアな気持ちになれている気もします。

今後、俳優として表現の幅を広げていくときに、自分に課している課題や目標はありますか？

M　映画が大好きなので、たくさん映画に出たいというのはあるんですけど、それ以上に、身を削って、誠心誠意向き合える役や作品に出会い、我が子のように愛せる作品を育てていきたいです。

蓄積された感情たちが、映画として形に残っていくのが楽しみですね。

M　はい、本当にそう思います。私は特に不器用なので、いろんな役を同時にということは今後も難しいというか、できないと思うので、一つひとつ大事に向き合いたいなと思います。

Did you ever have a period where you bottled up your emotions as a reaction to the difficulty of expressing your energy?

M　Yes, I did. I am incredibly clumsy with my emotions, and there were times when I felt excessive stress and completely shut down my ability to communicate. I couldn't control my energy, my mind would go blank, and sometimes I couldn't speak. Acting has been a very balanced job for me. While portraying a character requires delicate work, being able to express the emotions I've accumulated through acting leads to a sense of self-affirmation. It suits me, and I enjoy it.

You became independent in 2023. Was there a significant catalyst for this decision?

M　I never thought I would leave my agency, but in 2022, I attended the Busan International Film Festival. Up until then, I thought of film festivals as fun, festive events, but it turned out to be much more than that. It's a rare gathering of film professionals where you promote films and meet people you wouldn't encounter anywhere else. I went without any preparation and couldn't even speak English. I regretted not understanding what a film festival truly entailed. That experience triggered me to become independent and gain experience from scratch while I still could.

How has it been after spending six months as a freelancer?

M　It's been full of challenges. As a freelancer, there are things I can't do due to trust issues, the weight of responsibility, and many circumstances I wasn't aware of. I've become much more grateful to others, a hundred times more than before. On the other hand, receiving work directly has become even more rewarding, and it's brought out a sense of pure joy in me.

As you continue to expand your range as an actor, do you have any personal challenges or goals you've set for yourself?

M　I love movies, so I want to appear in many films. But more than that, I want to encounter roles and works that I can fully commit to with all my heart and soul. I want to cherish each project, finding and nurturing works I can love as if they were my children.

It must be exciting to see your accumulated emotions take shape as films.

M　Yes, I genuinely believe that. Since I'm particularly clumsy, handling multiple roles simultaneously will continue to be difficult. So, I want to cherish each one and face them with care.

松浦りょう　俳優　2014年映画『渇き。』で俳優デビュー。2019年に大河ドラマ『いだてん〜東京オリムピック噺〜』に出演し、2020年には映画『眠る虫』で初主演をつとめる。第27回釜山国際映画祭のキム・ジソク賞にノミネートされた2023年公開の映画『DECEMBER』では、17歳で殺人を犯し懲役20年を言い渡された加害者少女役を演じた。

Ryo Matsuura made her debut in the film *Thirst*. In 2019, she appeared in the historical drama *Idaten: Tokyo Orinpikku-banashi (A Tale of the Tokyo Olympics)*, and in 2020 played the lead role in the film *NemuruMushi (Sleeping Bugs)*. In the film *December* that was released in 2023 and nominated for the Kim Ji-seok Award at the 27th Pusan International Film Festival, Ryo played the role of a girl who committed murder at the age of 17 and was sentenced to 20 years in prison.

Treasures of the Wardrobe

ワードローブの宝物

ロクサーヌ・マゼリ

普遍的な服とは？　個性とは？　シンプルな洋服に程よくアクセントを加えるには？　ロクサーヌ・マゼリならそんな質問に的確なアドバイスをくれるだろう。約6年間アニエスベーのショップマネージャーを務め、1年前からはECサイトのスタイリングを手がける彼女は、アニエスベーの洋服を知り尽くした人物だ。パリのルード マルセイユのブティックに並ぶアイテム中心にタイムレスなおすすめを紹介してもらった。

　地味と表現するとネガティブに捉えられるかもしれませんが、私は好きなんです。それは、クリーンかつシンプルであることと同義。大切なのは素材と色の適切な組み合わせを見つけること。夏はリネンやコットン、冬はヴァージンウールやカシミアを選びます。マスキュリンなスタイリングが好きなので、夏はTシャツとジャケットでちょっと変わった合わせを楽しんだり、冬はオーバーサイズのニットをよく着たりします。アニエスベーのアイテムはまさに、私の好みにぴったりなんです。

　ジャンプスーツはアニエスベーの代表作。一度着てみれば納得できるはず。パターンや素材などすべてが巧妙。宝物を集めるように何年もかけて色違いや素材違いをクローゼットに納めています。

　アニエスベーのワークウェアの象徴でもあるカントン ジャケットも大好きで、毎回のコレクションに欠かせないアイテムです。落ち着いたエレガントなジャケットは、スタイリングに独特のスタイルをもたらし、全体のシルエットもユニークなものにしてくれます。

　私が勤めているのは、パリの中心地10区のルード マルセイユにあるブティックです。ここは10年以上前にお店ができるまでは、ギリシャ人一家が経営するスポンジ工場だったんです。昔ながらのファサードと当時の魅力はそのまま。ウッドのパネルやアンティーク家具、あたたかみのある色調の温室など異国を彷彿とさせる雰囲気もあるんです。

　アニエスベーのお店にはじめて足を運んだ時、衝撃を受けたのがアクセサリーでした。そして、シンプルなバンダナで洋服に捻りを加えるというアイデアがとても気に入りました。どんなスタイルでもバンダナひとつで見せ方を変えられるのがおもしろいんです。私の4歳と8歳の子どもたちも真似するんですよ。こういったアイテムの取り入れ方もアニエスベーの精神のひとつだと思います。

　私は何よりもまずひとりの女性としてアニエス・トゥルブレを尊敬しています。国際的に有名なデザイナー、アートコレクター、そしてアーティストのパトロンでもありますが、何よりもトレンドに流されず自分のスタイルを貫くことに成功しています。上質な素材を用いたタイムレスなスタイルも、文化や芸術に対する貢献も、彼女のDNAが脈々と受け継がれることで、時代を代表する女性になったんだと感じます。

　すべてのアイテムに時代を超えたストーリーがあり、洋服だけでなく、ギャラリーの運営やタラ号プロジェクトまですべてに一貫して彼女のメッセージを感じられます。それらの前衛的なメッセージは、フランスらしくもあり、ブランド設立以来、彼女が守り続けてきたものでもあります。控えめで慎重な彼女の姿勢は常に私を惹きつけています。

How does a garment become universal? What makes a style unique? How do you spice up an outfit in a simple way? Roxane Maselli is the one who can give sound advice on these questions. Having worked as a shop manager for six years and a women's stylist for the website for one year, Roxane knows agnès b. inside out. In the rue de Marseille boutique, Roxane gave us her recommendations for some timeless pieces.

I like sobriety, which can often be perceived negatively. A clean, simple style is always effective. What's important is finding the right mix of materials and shades. I easily opt for linen and cotton in summer and virgin wool and cashmere in winter. I like masculine looks, so an offbeat jacket with a T-shirt in summer or a slightly oversized jumper in winter are an integral part of my dressing room. The agnès b. style is a real sanctuary for me!

The jumpsuit is the chef-d'oeuvre of the agnès b. collection. Everyone loves it: try it, and you'll be convinced. My dressing room carries my 'treasures', as I like to call the iconic pieces I've adapted over the years in several colours and materials.

I'm also very attached to my Canton jackets, which are emblems of the workwear agnès b. style. They're the must-have pieces in the collection. Sober and elegant, they always bring an inimitable style to a look, making every silhouette unique.

I work in a boutique in the heart of the 10th arrondissement, on rue de Marseille. Over 10 years ago, before the shop opened, the site was formerly a sponge factory run by a Greek family. We have kept its old-fashioned facade and charm. With wood panelling, antique furniture, and warm Moroccan colours, the shop is an invitation to travel.

The accessories struck me most when I first arrived at agnès b.! I immediately liked the idea that a simple bandana could give an outfit a twist—I love twisting looks with this little square. It's fascinating how a simple bandana can completely change the way any style looks. My children, aged four and eight have already adopted it! And that's also the spirit of agnès b.

For me, Agnès Troublé is first and foremost a woman, an internationally renowned French fashion designer, an collector and patron of the arts who has managed to impose her vision without bowing to the dictates of the press and the fashion world. Whether it's for her approach to timeless relaxed style, with its emphasis on quality materials and durability, or for her contribution to the cultural and artistic scene, it's a DNA that makes her one of France's leading designers and women of influence.

Behind each garment is a story of a timeless creation, but beyond clothing, a universal message is carried through the galleries she runs and the Tara project. It's an avant-garde message that only Agnès has defended since the creation of the brand, thanks in part to French know-how. It is her modest and discreet commitment that has always appealed to me.

クサーヌ・マゼリ　アニエスベー ショップマネージャー, EC ウィメンズ スタイリスト　パリのルード マルセイユにあるスポンジ工場を跡地にしたアニエスベーの店舗にてマネ
ージャーを6年間務め、近年ではアニエスベーのECサイトのスタイリングを手がけている。

Having served as a manager at the agnès b. store located in a former sponge factory on rue de Marseille in Paris for six years, Roxane Maselli now also handles styling for the e-commerce

Beyond Words

Photography: Yudai Kusano
Styling: Reina Ogawa Clarke
Hair & Make-up: Yuko Aika
Text: Tomoko Ogawa

agnès b. stories

KAHO

Interview

agnès b. stories

2004年に12歳でCMデビューし、今年で活動20周年を迎えた夏帆。
映画、テレビドラマ、舞台などメディアを横断しながら、近年は、ドラマ『silent』『ブラッシュアップライフ』などの
話題作で圧倒的な存在感を見せる彼女は、どのように、自分らしいスタイルを見つけていったのか。
体現者として、芝居を、人生を楽しもうとする彼女の今は、好奇心と発見で満ちている。

夏帆さんはどんな洋服を見たときに、ときめきますか?

K　今日の撮影中はどのスタイリングもとっても可愛くて、ずっとキュンとしていました。アニエスベーは、映画の衣装を手がけていたり、アーティストとコラボレーションをしていたり、カルチャーとの親和性が高いイメージがあったので、映画の中の登場人物になったような気持ちでした。あらゆるところに溶け込むというか、ベーシックだけど、それだけじゃなく、アイコニックで、遊びもあって、最近現場が続いていたので、久しぶりにおしゃれしたいなという気持ちになりました。

服をいざ買うというときに、決め手になるポイントは何でしょうか?

K　わりと直感で決めるタイプかもしれないです。例えば、似たような服ばかり持っているから、そうじゃないものを選んでみよう、と悩んだ末に買ったものは、結局着ない場合が多い気がして。なので、着たときに自分に馴染むものかどうかが基準かもしれません。

今回着用されたアイテムは、夏帆さんに馴染むものだったんですね。

K　いつもは赤と青だったら青を選んでしまうけれど、普段は選ばない色味だったり、サイズ感だったりがすごくしっくりきたんですよね。自分はこういう洋服も着られるのか、という嬉しい発見がありました。

夏帆さんが長く大切にしているものには、どんな共通点がありますか?

K　買うときと同じでやっぱり、身につけていてしっくりくると思えるもの、落ち着くもの。それが自分らしさと言えるかもしれないですね。洋服や物を選ぶとき、物件を探すとき、人間関係においても、結局はそういう直感みたいなものが信用できると思っているところがあります。それに、気に入ったものは毎日着たくなってしまうので、長く着るために大切にしなくちゃと愛着がさらに湧いてくるものですよね。しっくりくるという理由以外でも、何かのときに着ていた服だったり、すごくがんばって買ったものだったり、そういう思い入れがあるものも、大切にしたいです。

日頃から俳優としてセリフと向き合っている夏帆さんは、とても真摯に言葉と向き合っている印象があります。

K　私は考えていることや感じていることを言葉にして相手に伝えるということが、昔からすごく苦手で。俳優という仕事は、人の書いた言葉をたくさん話しますが、自分の言葉で表現する機会はあまりないんです。だからこそ、説明しきれないし言語化はできないけど、そこにある何かみたいなものがとても面白いし、うまく言葉にできないものを表現したいと思いながらいつもお芝居をしています。とはいえ、作品をつくりあげる上で、一つひとつスタッフの方や共演者の方とコミュニケーションを取ることは大事なので、言葉にして伝えることは、私にとって目下の課題ですね。選ぶ言葉ひとつでも伝わり方は変わってしまいますし、「あー、日本語って難しい!」と日々悩まされますが、できないからと投げ出さずに、自分なりに考えて、ちょっとずつでも言語化できるようになりたいです。

20代からご自身で作品を選択するようになったそうですが、取捨選択を重ねてきた夏帆さんが、改めて自分について最近気づいたことはありますか?

K　自分の中にある何かを引き出してお芝居できる人はすごく素敵ですし、羨ましいと感じる反面、私は内側から湧き出る何かを表現したい、伝えたいという思いがあまりなくて。表現者というタイプではないし、向いてもいないのに、この仕事をしていていいのだろうか、とずっと考えていました。でも、何が楽しくて、何をやりがいとしてこの仕事を続けているかを振り返ったときに、自分は体験することが好きなんだ、と気づいて腑に落ちたんです。役者はいろんな場所に行けるし、たくさんの人を演じて理解していく中で、他人の人生を擬似体験できる。それが私にはすごく面白いんだなって。芝居をしていると、知らなかったさまざまな自分にも会えますし。たぶん、知らない何かを学ぶことが好きではあるけれど、本から得た知識としてというよりも、私は実際に自分のからだとこころを使って体験したいんだとわかってから、仕事が以前にも増して楽しくなりました。そういう話をしていたときに、友達がいいことを言ってくれて。「夏帆は表現者じゃなくて、体現者なんだね」って。それを聞いたときに、私もこの仕事をしていてもいいんだと思えたんですよね。

それこそ天職だと思います。現在30代の夏帆さんですが、これから40代に向けての指針について教えてください。

K　先のことはわからないのであまり決め込まない方ではあるのですが、健康だけは気をつけようと思っています。そうすることで、その時々、自分の好奇心が向かうもの、気になることや面白がれるものに正直でいられるでしょうし、そういう感覚を大事にしていきたいです。

Kaho made her debut in 2004 at the age of 12, and this year marks the 20th anniversary of her career.
Kaho has been active across media, including films, TV dramas and theatre, and in recent years, has shown an overwhelming presence in high-profile productions such as the dramas *Silent* and *Brush Up Life*.
Kaho is full of curiosity and discovery as she tries to enjoy theatre and life.

What kind of clothes spark joy in you?

K   Every style was so cute in today's shoot, I felt a constant thrill. agnès b. has always given me the impression of being closely tied to culture, whether designing costumes for films or collaborating with artists. It made me feel like a character in a movie. The clothes blend into any setting; they're essential yet iconic, with a playful touch. I've been on set a lot recently, so it made me want to dress up again after a long time.

What are the deciding factors when you buy clothes?

K   I am the type who decides quite intuitively. For example, if I think, 'I already have a lot of similar clothing, so I'll choose something different,' I often end up not wearing them. So, the criterion might be whether the clothes feel right and suit me when I wear them.

The items you are wearing this time suit you very well.

K   Usually, if I had to choose between red and blue, I would pick blue, but this time, the colours and sizes I wouldn't normally choose felt just right. It was a delightful discovery to realise that I could wear this kind of clothes, too.

What common traits do the things you have treasured for a long time have?

K   Just like when I buy something, it has to feel right and comfortable when I wear it. That might be what I consider my true self. Whether it's choosing clothes or items, looking for a place to live, or in relationships, intuition is something you should ultimately trust. Also, when you really like something, you want to wear it every day, so you naturally take good care of it to make it last longer, and your attachment to it grows. Besides feeling right, I also cherish things with special memories, like clothes I wore at significant moments or items I worked hard to purchase.

As an actor who constantly engages with scripts, you seem to have a very sincere relationship with words.

K   I've always needed to improve putting my thoughts and feelings into words and conveying them to others. As an actor, I speak many words written by others, but I don't often have the chance to express myself in my own words. That's why I'm fascinated by the ineffable elements that can't be fully explained or verbalised. I always strive to express things that are difficult to put into words through acting. However, communicating with each staff member and co-actor is crucial in creating together, so expressing myself in words is a significant challenge. How a message is received can change based on your chosen words. Every day I struggle with how difficult the Japanese language is! But instead of giving up, I want to think things through my own way and gradually become better at verbalising my thoughts.

Since your twenties, you've started choosing your projects yourself. Is there anything you've recently realised about yourself from all the choices and decisions you've made?

K   I admire and envy people who can bring out something from within themselves and express it through acting. On the other hand, I don't have a strong desire to express or convey something that wells up from within me. I've often wondered if it's okay for me to be in this profession, given that I'm not the expressive type and don't feel suited to it. However, when I reflected on what makes this job enjoyable and fulfilling for me, I realised that I love experiencing things. Being an actor allows me to go to different places and by playing various characters, I can experience other people's lives vicariously. That's something I find incredibly fascinating. Acting also allows me to discover unknown aspects of myself. I probably enjoy learning about things I didn't know, but rather than just acquiring knowledge from books, I prefer to experience it with my body and mind. After realising this, my work has become even more enjoyable than before. When I talked about this with a friend, they said something insightful. They told me, 'Kaho, you're not an expresser; you embody.' Hearing that made me feel like it was okay for me to continue in this profession.

You were born to be an actor. You are in your thirties now; could you tell us your guiding principles as you move towards your forties?

K   I don't usually make rigid plans since I can't predict the future, but I want to be careful of my health. By doing so, I can stay true to whatever piques my curiosity at any given time and to things that catch my interest or seem fun. I want to value those feelings and maintain that sense of wonder.

夏帆　俳優　2007年『天然コケッコー』で映画デビュー。第31回日本アカデミー賞新人俳優賞をはじめ、新人賞を総なめにする。以降、映画、ドラマを中心に活躍。最新作に、2024年3月放送開始の夜ドラ『ユーミンストーリーズ』、ヤマシタトモコによる同名マンガは瀬田なつき監督が実写化した映画『違国日記』（2024年6月公開）などがある。

Kaho debuted in the film *Tennen Kokkekō* in 2007. She was awarded numerous newcomer awards, including the Newcomer of the Year at the 31st Japan Academy Prize. Since then, she has been active primarily in films and dramas. Kaho appeared in the drama *Yuming Tribute Stories* airing in March 2024, and the film adaptation of Tomoko Yamashita's manga *Ikoku Nikki (Different Country Diary)*, directed by Natsuki Seta, which was released nationwide in June 2024.

チャド・ムーアがアニエス・トゥルブレの名前をはじめて知ったのは、写真家 ライアン・マッギンレーのインタビューを読んでいる時だった。そこにはライアンがパリでアニエスと一緒に食事を楽しんだという思い出が語られていた。やがてチャドはフロリダからニューヨークに移り住み、ライアンのアシスタントとして仕事をスタート。2014年、チャドはアニエスがニューヨークにいると聞き、彼女のホテルに写真集を数冊届けた。驚いたことに、数週間後アニエスから連絡があり、ニューヨークにあるアニエスベーのギャラリーで個展を開催することが決まったのだ。同年、パリのギャラリー デュ ジュールでも個展を行い、青山のアニエスベー ギャラリー ブティックと渋谷のアニエスベー カフェでも個展を開くことになった。その時の思い出がきっかけでチャドは東京を「地球上で最も好きな場所」と呼ぶ。そんなチャドが、東京とアニエスベー、そしてアニエスとの思い出を当時の未公開写真と共に綴る。

Dear Agnès Troublé

ふたりの深い友情と愛

チャド・ムーア

　2019年、青山に新しくできたアニエスベー ギャラリー ブティックで個展を開催した。東京のアニエスベーのギャラリーはかなり広かったので、写真が大きくプリントされているのを見てとても興奮したのを覚えているよ。さらにエキサイティングだったのは、世界中から僕の友人がオープニングのために集まってくれたこと。その後、数日間は本当に楽しかったよ。インスタレーションを行ったり、インタビューを受けたり、ギャラリーの中で飲んで、踊って。それから、渋谷にあるロックが聞けるバー、グランドファーザーズやBeat Cafe、カラオケやルーフトップバーにも行ったよ。とにかく東京中を走り回った。オープニングは金曜日だったんだけど、お客さんの入りがすごくて、500人くらいは来てくれたかな。こんなに楽しい思い出は他にないよ。

　2022年5月、写真集『Anybody Anyway』のサイン会のために再び東京を訪れた。アニエスベーチームの提案で、アニエスベー渋谷店の壁面に僕の写真をコラージュすることになったんだ。今までで一番気に入っているインスタレーションに仕上がった。オープニングパーティーでは、2019年に行った個展で出会った人たちにも再会することができた。こんなにも住む場所が離れている人たちとまた会うことができるのは本当に嬉しかった。

　日本はとにかく最高だ、と訪れるたびに思うよ。東京で遊んだ友達はみんなそれぞれが住む場所へと戻って行った。楽しい思い出が多いほど離れると寂しくて、お互いに飛行機の中から泣き顔のセルフィーを送り合ったよ。一生に一度しかないような感覚だった。

　最後にアニエスとの思い出を……。僕は彼女の仕事に対するまったく妥協がない姿勢を本当に尊敬しているし大好きだと感じる。彼女は働いてない瞬間がない。本を執筆しながら、新しい個展のキュレーションをしながら、チェット・ベイカーの話をしながらコレクションの制作に取り組む。僕はそんな彼女のエネルギーに一瞬にして虜になってしまったんだ。時折、彼女のオフィスの屋上で夕日を眺めながらお酒を飲んだ時のことを思いだす。彼女は屋上を歩き回って、ハート型の石を見つけてプレゼントしてくれた。そして、その石がきっかけでこの建物を自分のスタジオにしたのだと教えてくれた。

　アニエスは僕の写真家としてのキャリアだけでなく、人生においてもとても大切な人物。彼女の人や世界を見る視点からは学ぶべきものがある。彼女がアートやアーティストを尊重する姿勢にはいつでも好感が持てる。彼女は作品を投資目的で買おうなんて考えもしない。バスキアであろうと、無名のアーティストであろうと、それぞれの作品が持つストーリーを愛している。彼女は本当にアートを信じているし、それは稀有なことだと感じる。気取ったところがなく、ただ愛と驚きと興奮に満ちている。そんな彼女が動いているところを見れること自体がギフトなんだよ……。誰も見向きもしないようなものをiPhoneで撮っている彼女の姿を見るのが大好きなんだ。

　アニエスベーと日本について、そして僕が両者に対して抱く愛について、書きたいことは山ほどあるけれど、分厚い本になってしまいそうなのでこの辺にしておくよ……（笑）。アニエス、おめでとう。これからもたくさんの冒険を！

Chad Moore's first encounter with the name of Agnès happened while reading interviews of photographer Ryan McGinley, who shared his bon-vivant-esque stories of hanging out with her in Paris. Eventually, Chad moved to NYC from Florida and ended up working as Ryan's assistant. In 2014, Chad heard that Agnès was in town and dropped a few of his photo-books at her hotel. To his great surprise, Agnès reached out to him several weeks later, and he was offered a show at the agnès b. store in New York. Another show at Galerie du jour in Paris followed the same year, as well as two exhibitions at agnès b. galerie boutique in Aoyama, and at their café in Shibuya, Tokyo. The city became his 'favourite place on Earth'. Chad shares previously unpublished images of his time in Tokyo, from exhibition openings to partying with friends, as well as his memories of meeting Agnès and forging a long-lasting relationship ever since.

In 2019, I held an exhibition at the newly opened agnès b. galerie boutique in Aoyama. The gallery in Tokyo is quite massive so it was super exciting to see some of the work printed so large. Even more exciting was that a bunch of my friends from around the world were coming out for the opening. The next few days were insane; installing and interviews and drinking and dancing in the gallery and just all of us running around Tokyo going from bar Grandfathers to BEAT café to karaoke to hanging on some rooftops and on and on. The opening was a Friday and the turnout was crazy, maybe 500 people came through. I've never had more fun.

In May 2022, I was in Tokyo for a signing of my book *Anybody Anyway*. Since I was here, the agnès b. team suggested we did a collage of my images on one side of the entire Shibuya store. It was one of my favourite installations ever. At the opening party, I got to see a lot of people who came to the previous show in 2019. It's so fun to live a world apart, and to reconnect once again.

My visits to Japan were such epic trips. My friends and I all had to fly home separately and we all sent crying selfies to each other from the plane. It was one of those once in a lifetime feelings.

I would like to conclude with memories of Agnès. One of my favourite things, the most admirable thing about her is her insane work ethic, she's never not working. She will be working on the collection while looking through a book, curating a new exhibition, all while telling you a story about Chet Baker. I was instantly in love with her energy. One time, we ended up having drinks on her roof while watching the sunset over Paris. Agnès would walk around and find heart shaped rocks to give to me, telling me this was the reason why she decided on this building as her studio.

Agnès has been a seminal figure in not only my career, but my life. There is something to be learned from the way she looks at people and the world. I've always loved the way she respects art and artists, she's never thought about flipping pieces or buying as investments, she loves the stories behind the works, whether it be Basquiat or someone you've never heard of. She truly believes in it and I think that is the rarest thing. There is nothing pretentious, it's just full of love and wonder and excitement. Just watching her operate is a gift... I love watching her constantly take iPhone photos of things you wouldn't look twice at.

There's so much to write about agnès b. and Japan, and all the love I have for both, but it might end up a very big book. Congratulations, Agnès. To many more adventures to come!

チャド・ムーア　写真家　ライアン・マッギンレーの元でアシスタントをつとめた後、独立。ニューヨークのダウンタウンアートの次世代を担う。2016年に発刊した写真集『Bridge of Sighs』が話題となる。アムステルダムやアントワープ、パリ、東京など世界各地で展覧会を開催。現在は主にファッションの分野で活躍している。

Chad Moore worked as an assistant under Ryan McGinley before going independent. Part of the next generation of New York's downtown art scene, his 2016 photographic book *Bridge of Sighs* has been much talked about. Chad has exhibited internationally in Amsterdam, Antwerp, Paris and Tokyo. Chad currently works mainly in the field of fashion.

2023年に、青山のアニエスベー ギャラリー ブティックで個展を開催した画家の水戸部七絵。
「座る人 "Sit-in"」というタイトルが付けられた本作は、留学先のウィーンで目にした沖縄の座り込み
運動の報道がきっかけとなり、抗議をする者の絶対に動きたくない強い意志の表れが表現されている。
様々な種類のレコードに描かれた肖像画にはどのような想いが込められているのだろうか。
千葉県郊外の、のびやかな田んぼの中に佇むアトリエ。創作の原点である画材やレコード、
過去の作品に囲まれ、インスピレーションの源が垣間見える環境に立つ彼女は、
作品から溢れ出る力強さとは異なり、柔らかくほがらかに笑う様子が印象的だった。

キャンバスからはみ出して
水戸部七絵（画家）

Beyond the Canvas
NANAE MITOBE (Painter)

Photography: Kiyoe Ozawa
Text: Chikei Hara

アニエスベーにはどのような印象を持たれていましたか？

M　アニエスさん自身がアートのビッグコレクターでもあるので、そのコレクションの数々が素晴らしく、尊敬しています。以前パリにあるラ・ファブを訪れたときにも、私一人しか観客がいない状況で広大な空間でパフォーマンスが繰り広げられる贅沢な時間を体験しました。また彼女の「ロックンロールは不滅」というかっこいいメッセージに強く共感していて。私がやっていることにもどこか根幹はつながっているのではないかと思っていて、それを受け入れてくれるのもアニエスベーらしさだと感じました。

今回展示される「座る人 "Sit-In"」はどのようなシリーズなのでしょうか？

M　2022年に留学していたウィーンで目にした辺野古の座り込み運動のニュースがきっかけになっています。海外で生活しながら人種や社会問題について考えていた頃に、こんな形で日本に触れることもあるんだと思いました。もともとウィーンに行くきっかけになったのも、当初予定していた滞在地が戦争の影響で渡航困難になってしまったからでした。この地ではロシアや内戦から逃れた移民と出会って様々なお話を聞く機会が多く、世界には戦争が常に起きてしまう地平線があることを肌で感じました。

座り込み運動は非暴力を通して平和的な解決を示す態度とも言えます。水戸部さんはどのように捉えたのでしょうか？

M　実は私自身が神奈川県の座間市出身で米軍基地を近い存在として感じながら過ごしてきたルーツがあり、沖縄の辺野

古に住んでいる人も状況はそこまで遠くない気がしています。街中にいる米軍の方とカジュアルな距離感で話すことも、小学校の頃から飛行機の訓練を見ることも日常の一部でした。日本にいると国境を意識するきっかけが少なく、そのような環境にいても遠い存在に捉えてしまいますが、心のどこかで起きないものと思っていた戦争が現実に行われていることに大きなショックを受けています。

遠くの存在や距離に対する思考をどのように捉えていますか？

M　なぜ顔を描いているのかという理由にもつながると思いますが、自分が有色人種であることや美を対象とした憧れ、日本人としてのコンプレックスを抱くことは一体どういうことなのかを考えています。コロナが落ち着き、留学を経験したことで、これまでよりも密接なコミュニケーションが取れて、人々との距離感が縮まった状態で絵を描くことができました。距離の遠い存在や象徴的な憧れをモチーフとして普段作品を描いているので、ウィーンでの経験を通して新たなテーマが芽生え、価値観をアップデートできたことは、個人的に興味深いポイントです。

政治的な態度や美術作品などの同時代性をどのように捉えていますか？

M　昔のことを同じように繰り返し手を動かすことに、私は特別な意味を成さないと思っています。常に新しくあることは一つの習性でもあって、こうするべきだという規範に従うことなく、物事を別のものとして捉える変換のあり方を開拓していくことに面白さを感じています。絵画に限らず芸術で扱われなかった素材や認められてこなかったことを認めさせる行為

をしている作家をリスペクトしていて、ジャンルに囚われずたくさんのことが更新されていく現代を見ていると、画家として絵画にしかできないはみ出し方を模索してみようと思えます。

この作品（P120左）にレコードが組み込まれているのはどうしてですか？

M　レコードを扱ったシリーズをパブリックに発表するのは今回が初めてなのでどこか不思議な存在として感じています。レコードに自分がペインティングすることを一種のコラボレーション的な感覚で捉えて、この「座る人」という個展では、ファッション、音楽、グラフィティ、そして抗議する人という条件が揃うことで、ヒップホップというジャンルが成立する空間を作り上げようとしています。そこにDJが来て、観客がダンスをすれば成立するでしょう。レコードに施したペインティングは、その重要な役割（グラフィティ）を担っています。何をもってその音楽であるかという定義付けは意外と人によって様々で、私がこれまで続けてきた肖像画における人種や顔についての問題ともルーツがあると感じています。

作品を作る上で意識の変化もありましたか？

M　私が特段音楽に精通しているわけでもないので、描いている最中は意外とレコードの画面に意識を引っ張られることはなくて。音楽に詳しい方が見る面白さもあると思い、歴史的に価値がある物からガラクタになってしまったものまで様々なジャンルを横断しながら選んでいます。ですが私の中ではレコードの印刷物としての性質も重要で、音楽の商業的な資本行動としての側面を考えながら、レコードとして刷られてしまった存在をマーケットに参入してしまう芸術や絵画の比喩として扱っています。

タイル状に組み合わさったレコードに肖像画が描かれることに面白さを感じました。

M　ヒップホップというジャンルでDJが音楽をミックスさせながら構築していくスタイルがあるように、画家としてビジュアルをいろいろミックスさせることで作品を成立させたいと思いました。絵画には原理的に四角の世界を描くという制約があって、31cm角の正方形のレコードが画面を細分化する状況は構成としても面白いなと思っています。今まで四角を通して考えてきたことが、画面が分割されたり四角からはみ出て人型の作品が生まれたりすることで、新たな可能性を感じました。レコードも絵も印刷されてしまうことで消費と結びついてしまう構造の問題をはらんでいて、結果的に世の中にたくさん広がっていると思うけど、異なる形で捉えられる気もするんですよね。子供の頃、親に買い与えられたお洋服や本などに落書きする行為をずっとしていて、そうすることには自分だけのものにするマーキング的な意味もあったりして、宝物みたいに大切に思える気持ちが芽生えました。レコードの音楽的な魅力やジャケットの良さも大衆的であるからこそ、絵を通してマーキングする面白さがある気がしています。

水戸部さんが肖像画を描き始めたきっかけは？

M　ものすごく遠い存在だからこそ、肖像を描けるようになったきっかけにマイケル・ジャクソンがいます。私は全くかけ離れているものや知らないものに興味を持つことが多くて、芸術的な存在として認識しているマイケルの美しさを抽出する作業がペインティングなら出来るのではないかと思って、これまで何度も肖像画を描きました。マイケル自身が整形したり、絶頂期がある中で崩壊するさま、亡くなってからスキャンダラスな報道が行われてしまうことに、美術で言うところの遺跡的な亡びの美しさが重なるように思いました。私にとっては社会的に何か行為を起こした人たちや、人生の浮き沈みがあって振れ幅が大きい人の方が、その人の背景をいかに描くこ

とができるか想像を掻き立てられます。今回のシリーズでも、モハメド・アリなどヒップホップに関わる意義を感じられる人を描いています。

完璧ではなく崩れているさまをどのように捉えていますか？

M　例えば美術館で腕が折れたり鼻がもげている石膏像やヴィーナスを目の当たりにしたときに、そのイメージが築かれた原型となる姿形があることや過去に色があったことを知ることで、滅びていくさまに美しさを感じてしまう。そこに自分の美意識はある気がしています。絵を描くという行為はどうしても、物として崩れていき、さらに手前からある行為を模索することでもあります。構図を完璧に取って絵画を構成するとただの心地のいい絵になってしまうことに違和感を感じていて、そういう絵には私が発表する意味が伴わないと感じたときに、描き崩すこと、過剰に描くことを通して美術に関わろうと思えました。

水戸部さんはどのような感覚で絵の具の質感や色彩を捉えていますか？

M　小学生の頃にゴッホの油絵を見て感動した体験が、昨日見た記憶のように脳裏にあって。油絵の質感や、発色の美しさがいまだに脳内にインプットされています。そうした経験から私自身は絵画原理主義のような考えを持っているので、絵の具以外を使ったものは絵画ではないというルールを自分の中で設定して、一貫した制作を行っています。今回は四角の領域からはみ出したものが初めて登場するので、自分のルールから逸脱するリスキーな挑戦でもあると感じています。

アニエスベーでの展示をどのように考えられましたか？

M　アニエスベー ギャラリー ブティックでの展示のお話をいただいたときに、この場所が音楽やロックという固有のカテゴリーに囚われることなく複合的にファッションやアートと絡み合う空間であると感じました。アニエス自身のロックなスタンスや精神を表した店内には存在感のある素敵な赤いソファがあり、彼女が残した多くのメッセージやアフリカへのリスペクトも感じられる空間で面白いなと思って、私自身が考えていた座る人とヒップホップ、ファッションという要素のミックスはアニエスベーとのコラボレーションだからこそ存在しています。

作品があることでいろんな人に見られる状況が生まれ、私たちが展示を通してこの問題に参加して考えられるのかなと思います。

M　アートと音楽やファッションを絡める際に、表面上はキャッチーでポップになってしまうので親しみやすい空間になることはあるけど、そこから背景や裏側、歴史のルーツの深さを知ることで問題に参加する入り口になるといいなと思っています。

水戸部さんにとって、絵を描くことでしか表現できないことって何ですか？

M　難しいけど本質的な質問でもありますね。絵画に全く触れていない人からしたらピンとこないかもしれないけど、絵が輝いて見えたり絵に入り込むような瞬間があって、それだけで全てを把握できたような感覚になることが絵画を通して起こります。その経験を一度してしまったからには、もう出られなくなることもあります。平面作品と大きくカテゴライズされることはあるけど、その中のとても狭い専門的なジャンルだと思っていて、それを本質的なものとして捉えられる人はいまの時代には少ない気がします。私自身は絵画の良さを知っているからこそ、それを伝える宿命があるように勝手に感じています。

In 2023, painter Nanae Mitobe held a solo exhibition at the agnès b. galerie boutique in Aoyama, Tokyo. Her work titled *Sit-In* was inspired by news coverage of the sit-in protests in Okinawa while studying abroad in Vienna. This piece reflects the protesters' strong will to remain unmoved. What kind of emotions are conveyed through the portraits painted on various types of records?

What is your impression of agnès b.?

M   I greatly respect Agnès as a significant art collector, and her collection is truly remarkable. When I visited La Fab. in Paris, I experienced an extraordinary moment witnessing a performance unfolding in a vast space where I was the only visitor. I also strongly resonate with her cool message, 'Rock'n'roll is not dead'. I believe there is a fundamental connection to what I do, and I feel that agnès b. embraces that connection, which is very characteristic of the brand.

Please tell us about the *Sit-In* series that was exhibited this time.

M   The series *Sit-In* was inspired by news coverage of the sit-in protests in Henoko that I saw in the news while studying in Vienna in 2022. It was a moment when I realised that even while living overseas and thinking about racial and social issues, I could still connect with Japan in such a way. My decision to go to Vienna was influenced by the fact that my original planned destination became inaccessible due to war. In Vienna, I met many immigrants who had fled from Russia and civil wars and had the opportunity to hear their stories. It made me acutely aware that there are always horizons where wars are ongoing in the world.

How do you perceive the sit-in movement, which can be seen as an attitude demonstrating a peaceful resolution through non-violence?

M   I grew up near a US military base in Zama City, Kanagawa Prefecture. Because of that, I feel connected to the people living in Henoko, Okinawa. From a young age, I was used to seeing and interacting with people from the US military around town and watching flight training exercises as part of everyday life. We don't often think about national borders in Japan—they seem distant even in this environment. The reality of war, something I thought could never happen, has shocked me.

How do you perceive the distant existence of places and people?

M   I often think about what it means to be a person of colour, to admire beauty, and to deal with my insecurities as a Japanese person. After the pandemic settled down and having the experience of studying abroad, I had the chance to communicate more closely with people and feel a lot more connected while creating my art. I usually use distant entities and symbolic idols as motifs in my work, so the experiences in Vienna sparked new themes and allowed me to update my values, which I find fascinating.

How do you perceive the contemporaneity of political attitudes and artworks?

M   Repeating actions the same way as in the past doesn't hold any special meaning for me. Being constantly innovative can also be a habit, and I find it intriguing to explore ways of transforming and perceiving things differently without adhering to conventional norms. I respect artists who use materials or themes not traditionally recognised in art. Seeing how so many things are evolving and getting updated in today's world, regardless of genre, makes me want to push boundaries as a painter.

Why did you incorporate records into this work (P120 top left)?

M　This is my first time publicly presenting a series involving records, and it feels somewhat mysterious. I see painting on records as a kind of collaboration. I'm trying to create a space where hip-hop comes to life through the combination of fashion, music, graffiti, and protest. The space will genuinely embody hip-hop if a DJ comes in and the audience starts dancing. The paintings on the records play an essential role as graffiti in this set-up. The definition of what makes that music can vary greatly from person to person, and I feel it's connected to the issues of race and identity in the portraits I've been working on.

Have there been any changes in your awareness while creating these works?

M　Since I am not especially familiar with music, my focus wasn't overly drawn to the records themselves while painting. I chose records across a wide range of genres, from those with historical value to those considered junk, which might be interesting for people knowledgeable about music. However, for me, the printed nature of the records is also important. I thought about the commercial aspect of music production and used records as a metaphor for how art and painting, once created, inevitably enter the market.

It's interesting to see portraits painted on records arranged like tiles.

M　Just as DJs create something new by mixing music, I wanted to establish my work by combining various visual elements as a painter. There is an inherent challenge in painting within the rectangular confines of traditional canvases, and I found it fascinating to use 31cm square records to subdivide the visual space. This method of breaking up the canvas or extending beyond the square frame to create human-shaped works opened up new possibilities for me. Both records and paintings, once printed, become part of a consumer structure and are widely disseminated, but they can be perceived differently. As a child, I used to doodle on clothes and books given to me by my parents, marking them mine and making them feel like treasured possessions. The popular appeal of records, with their musical allure and interesting covers, makes it fun to mark them with my art. This act of painting on records brings a unique layer of personal and artistic significance, similar to how doodling on my childhood belongings made them feel uniquely mine.

What inspired you to start painting portraits?

M　Michael Jackson was the catalyst for painting portraits for me precisely because he felt so distant. I am often drawn to things that are vastly different from or unknown to me, and painting could be a way to extract and capture the artistic beauty I saw in Michael. His journey—from undergoing numerous plastic surgeries and experiencing a peak followed by a dramatic decline to the scandalous reports after his death—paralleled the concept of the decayed beauty found in ruins in art. Individuals who have made significant social impacts or experienced extreme highs and lows inspire a desire to imagine and depict their backgrounds. In this series, for instance, I have painted figures like Muhammad Ali, who are relevant to hip-hop culture.

How do you perceive the beauty of something that is not perfect but decayed?

M　When I see plaster statues and sculptures in museums with broken arms, I am struck by the beauty in their decay, knowing that these forms once had an original, intact state and perhaps even colours in the past. My sense of aesthetics is rooted in this appreciation of the beauty in ruin. If I were to create a painting with a perfect composition, it would become a nice but ordinary image, which feels dissonant to me. Presenting such conventional paintings lacked significance, so I turned to deconstructing and excessively painting to engage with art in a meaningful way.

How do you perceive and capture the texture and colour of paint?

M　I vividly remember being moved by Van Gogh's oil paintings when I was in primary school, and the texture and vibrant colours of oil paintings are still deeply imprinted in my mind. Because of this experience, I hold a somewhat purist view of painting, believing that anything other than paint does not qualify as painting. I have set this rule for myself and have consistently adhered to it in my work. This time, however, I am introducing elements that extend beyond the rectangular frame for the first time, which feels like a risky challenge that deviates from my usual rules.

How did you approach the exhibition at agnès b.?

M　When I was offered the opportunity to hold an exhibition, I felt that this space is one where fashion and art intertwine, without being confined to specific categories like music or rock. The store's interior, reflecting Agnès' rock spirit and stance, features a striking red sofa, numerous messages from her, and a sense of her appreciation for Africa. This mix intrigued me, and I realised that my concept of combining the themes of sit-ins, hip-hop, and fashion aligns perfectly with this collaboration with agnès b.

Having the artwork on display allows various people to see it, which allows us to engage with and think about this issue through the exhibition.

M　When integrating art with music and fashion, the result can often be catchy and pop on the surface, creating a friendly and approachable space. However, I hope that by presenting the background, deeper layers, and historical roots, the exhibition can serve as an entry point for deeper reflection and engagement.

What can you express through painting that you can't through any other medium?

M　It might not be immediately apparent for someone who has never painted, but there are moments when a painting shines or draws you in, allowing you to understand everything through it. Once you've experienced that, it's hard to let go. While paintings are often broadly categorised as a two-dimensional art form, it's a very narrow, specialised genre within that category. Nowadays, there seem to be fewer people who can appreciate its essence. Because I understand the beauty of painting, I feel an inherent responsibility to communicate its value.

水戸部七絵　画家　以前より象徴的な人物の存在を対象として描いていたが、2014年のアメリカでの滞在制作をきっかけに、極めて抽象性の高い匿名の顔を描いた『DEPTH』シリーズを制作。2016年愛知県美術館での個展にて発表した。また、2020年に愛知県美術館に『I am a yellow』が収蔵された。

Nanae Mitobe previously depicted iconic figures as her subjects, but during her residency in the United States in 2014, she began creating the highly abstract DEPTH series, featuring anonymous faces. This series was exhibited in a solo exhibition at the Aichi Prefectural Museum of Art in 2016. Her work 'I am a yellow' was acquired by the Aichi Prefectural Museum of Art in 2020.

アーティストのヤビク・エンリケ・ユウジは、コラージュ制作を通して思いがけないものに美を見出してきた。ヴィンテージ雑誌の写真、日々のスケッチ、生活の中で拾った素材など、それらを無造作に並べることで彼の想像の扉は広がっていく。アニエス・トゥルブレも制作のインスピレーションを日常の中で見つけてきたアーティストのひとりだ。パリの街中で見つけたグラフィティ、空に浮かぶ雲の形、夕焼け……。両者はともにセレンディピティ（偶然の出会い）を大切に、表現を続けている。2023年のアニエスベー日本上陸40周年を記念し、ブランドのルックブックやアーカイブ写真を用いてヤビク・エンリケ・ユウジにコラージュ作品を制作してもらった。

Artist Yabiku Henrique Yudi has found beauty in unexpected things through his collage work. By arranging photos from vintage magazines, daily sketches, and materials picked up in everyday life, Yabiku expands the doors of his imagination. Agnès Troublé is another artist who finds inspiration in daily life. Graffiti found on the streets of Paris, the shapes of clouds in the sky, the colours of sunsets... Both artists cherish serendipity, the art of making fortunate discoveries by chance, and continue to express it in their work. To celebrate the 40th anniversary of agnès b. in Japan in 2023, Yabiku created a collage work using the brand's lookbooks and archival photos.

Doors of Imagination Opened by Chance

偶然が開く想像の扉
ヤビク・エンリケ・ユウジ

　ダンボールいっぱいのアニエスベーのアーカイブ。たくさんありすぎて選びきれないと感じたので、パッと見て直感で気になったものをセレクトしていきました。

　基本的にはすべて感覚で作っています。頭を真っ白にして、最初は何も考えずにはじめます。その素材が何に使われていたものか、どんな意味を持つのか。そうしたことはできるだけ忘れ、ひたすらレイヤーを重ねていきます。大きい作品や立体作品は今回制作したような平面作品より建設的に制作を進めますが、途中のちょっとしたエラーや好みでいろいろと変更するので結局はその時、その瞬間の気分、偶発性を大切に制作しています。

　アニエスさんは行く先々で石や木の葉などを集め、思い出として保管したり、デザインに使ったりするのが好きだと聞きました。僕も基本的にはどんなものもコラージュに適していると思っていて、日常的に鉄や紙、ゴミなどを拾い集めるのが好きです。その時の気分と構成にさえハマれば、どんな素材でも使うようにしています。

　アニエスベーと言えば、フランス人のフォトジャーナリスト、ピエール・レネ＝ウォルムスが撮影したバンド、ブロンディの写真がプリントされたアーティストTシャツを持っています。5年くらい前、仕事の打ち合わせの後にアニエスベーの青山店に立ち寄った際に、彼の写真展が開催されていて、その時のメインビジュアルがこのTシャツにもプリントされている写真だったので思わず購入してしまいました。コラージュで服を作ることに挑戦したいと思っているんですよ。

　ほとんどの作品は雑誌をめくって素材を探すことからはじめますが、どんな素材が見つかるかは最初の時点ではまったく想像できません。さまざまな切り抜きを重ねたり、また別の素材を組み合わせたりしていくと架空の世界観が生まれ、自然と物語が生まれます。作品の意味や内容が後からついてくるというのはコラージュという技法の独特な面白さだと感じていて、それを大事に制作しています。自分自身も制作過程や完成後にその意味に気づいたりすることもあるんです。僕の作品を見た人には、現実を忘れて自由にいろいろ想像しながら架空の世界を楽しんでもらえたら嬉しいです。

A box full of agnès b. archives. There were so many that it felt overwhelming, so I selected pieces based on my immediate intuition and what caught my eye at first glance.

I primarily create everything based on intuition. I start with a blank mind, not thinking about anything initially. I try to forget what the material was originally used for or what it means and simply focus on layering. For larger or three-dimensional works, I proceed in a more structured manner compared to the flat piece I created this time. However, I often make changes based on minor errors or my preferences along the way, so ultimately I value the mood of the moment and serendipity in my creative process.

I heard that Agnès likes to collect stones and leaves from her travels, keeping them as memories or using them in her designs. I also believe that almost anything can be suitable for a collage. I enjoy picking up iron, paper, and even trash in my daily life. I'm willing to use anything as long as the material fits the mood and composition of the moment.

Speaking of agnès b., I have an Artist T-shirt printed with a photo of the band Blondie, taken by French photojournalist Pierre René-Worms. About five years ago, I stopped by the agnès b. store in Aoyama after a work meeting, where the Pierre René-Worms' photo exhibition was being held. The main visual for the exhibition was the same photo printed

shirt, so I couldn't resist buying it. In the future, I
[ch]allenge myself by creating clothes using collage
[...]s.
[...] of my works start by flipping through magazines
[...] materials. By layering various cutouts and
[...] them with other materials, an imaginary

world takes shape, and a story naturally emerges. I
[...] it fascinating that the meaning and content of the w[...]
often come after the collage process, and I value tha[...]
creations. I hope that people who see my work can [...]
reality and freely imagine and enjoy the fictional wo[...]
emerges.

ヤビク・ユウジ　アーティスト　文化服装学院にて服飾を学んだ後、2017年よりコラージュを用いた表現活動を開始。2022年、YUKIKOMIZUTANIにて[...]
[...]HOUGHT」を開催。現在では、コラージュを中心にオブジェクトの作成や空間インスタレーションなど、その表現の裾野を広げている。

[...]g fashion at Bunka Fashion College, Yabiku Henrique Yudi began creating collages in 2017. In 2022, Yabiku held his largest solo exhibition, *AFTERTHO*[...]
[...]TANI Gallery. Yabiku is currently expanding his scope of expression to objects, spatial installations, and other forms, with collage as the focal point.

向き合って、自由になる
のん

Be Free

Photography: Naoto Usami
Styling: Mayu Takahashi
Hair & Make-up: Shie Kanno
Text: Tomoko Ogawa

NON

NHK連続テレビ小説『あまちゃん』の放送から10年以上が経ち、2023年に30歳の誕生日を迎えた、のん。
これまで俳優としてだけでなく、音楽、ファッション、映画、アートの世界で唯一無二のアーティストとして
ボーダレスに活動してきた彼女。「好き」を追いかけることで開いていく、未来について。

これまで何度かアニエスベーとコラボレーションしていますが、アニエスベーの洋服にどんな印象がありますか？

N　着ているだけで、映画に出てくる主人公みたいに情景やストーリーが浮かんでくる。それに、少年少女という感じではなくて、大人と子どもの間みたいな雰囲気があって、そういう意味でもキュンとくるというか、青春を感じます。

のんさんが表現するものも、青春感がありますよね。

N　青春感やキュンとくることが自分も大好きで、大事にしてるから、そういうところは共通するなと思います。それに、アニエスベーを着ていると、意思の強さは伝わるんだけれど、ギラギラしていない。街に溶け込んでいるんだけれど、ちゃんと強い人に見える気がします。

表現をするときに、原動力となっているものとは？

N　一番は、見てくれる人のリアクション、反応があること。「グサっと刺さりました」とか言われると嬉しいですし、人の感情を動かすことができるということが喜びです。それがあることで、次もがんばるぞと思えるし、何かつくったり、演じたりするときも、そんなふうに感じてもらえるものを目指したいと燃えますね。

それがどんな表現であっても、反応があることが嬉しいんですね。

N　そうですね。何でもよくて、「ゾッとした」とか「なんか気持ち悪かった」とかも嬉しい。これは、全部に共通していると言えるかもしれないですが、複雑な表現の方が好みなんですよね。例えば、笑ってるようにも見えるし、悲しそうにも見えるみたいなシーンが

あると、「よし、腕の見せどころだ！」という気持ちになる（笑）。もちろん、ハッキリした感情もあるんだけれど、なんかモヤモヤして、自分で自分の感情が定まらないときってあるじゃないですか。いくつもの感情が拮抗しているような瞬間をカメラに撮ってもらったり、そういう作品を残せたりすると、一番気持ちいいですね。見る人によって、違う理由で感情移入できるものが好きなんですよね。

30歳の誕生日に「Lula BOOKS」からリリースされた、アートブック『Non』で、変わらない自分勝手さを受け止められるようになったのが変化だと書かれていたのが印象的でした。素直に自分を受け入れられるようになったきっかけとは？

N　俳優として、脚本を読んで、キャラクターを解釈するときは、いい部分だけを見せるような表現は面白くないと考えられるんだけど、自分だけの表現に立ち返ると、その考えを排除してしまうところがちょっとあったんです。でも、のんになってから、作り手として自分自身を出さなきゃいけない。役名がついてない状態でいろんなものを表現しなきゃいけなくなって初めて、「いい子だよね」

だけで終わる人に見られたくないとか、弱そうに見られたくないとか、そういう欲が湧いてきて。そこから、うまくしゃべろうというよりは、うまくしゃべれないという不器用なところもちゃんと認めて、そこを表現していくのが自分に合ってるなと思うようになりました。作品、映画、ドラマ、バラエティに出ている人を見ていても、誰かと接しているときでも、そういうウィークポイントを見つけると、グッと入り込めるものだなと。だから、そういう部分を自分でちゃんと見据えようという気になりました。

これから30代の10年を、どのように楽しみたいと考えていますか？

N　歳を重ねるごとに、どんどん自由になっていく感覚はありますね。本当に未熟な部分やうまくできないことが自分にはあるけれど、経験を経ると、どんどん、「あ、こうやればいいんだ」という発見があるじゃないですか。そういう発見がもっともっとたくさん見つかると思うので、そういうのを全部生かして、自分の表現につなげられるなという予感がしていて、それが楽しみです。

More than 10 years have passed since the broadcast of NHK's TV series *Amachan*. In 2023, NON celebrated her 30th birthday. She has been active not only as an actor but also as an artist in the world of music, fashion, movies, and art, working across all genres. NON told us about the future that opens up as she pursues what she likes.

You've collaborated with agnès b. several times so far. What impression do you have of agnès b.'s clothing?

N　By wearing it, you can envision scenes and stories like a character in a movie. It has an atmosphere that's not quite childlike, but more like something between an adult and a child, making it feel nostalgic and giving a sense of youthfulness.

The things you express have a sense of youthfulness, don't they?

N　I love that youthful feeling and those heart-fluttering moments, and I cherish them, so that's something we have in common. Also, when I wear agnès b., it conveys a strong will without being flashy. It blends into the city, yet still makes you look like a strong person.

What is the driving force behind your expression?

N　The biggest driving force is the reaction and response from the audience. It makes me happy to hear someone say, 'That really hit me.' The joy of being able to move people's emotions is very fulfilling. It motivates me to keep trying my best, and whenever I create or act, it fuels my desire to aim for something that evokes those kinds of feelings.

So, you're happy to get a reaction no matter what kind of expression it is.

N　Yes, exactly. Anything is fine, like 'It gave me chills' or 'It felt a bit unsettling'—those reactions make me happy too. I prefer complex expressions. For example, a scene with someone laughing but also seeming sad at the same time feels like a challenge, but I would think, 'Alright, this is my chance to shine!' (Laughs) Of course, there are clear emotions too, but there are times when

you feel unsure and conflicted about your own feelings. Capturing moments where multiple emotions are at play and creating works that portray those moments is the most satisfying for me. I like pieces that allow different people to empathise for various reasons.

In your art book *Non*, released by Lula Books on your 30th birthday, you mentioned that you changed in your ability to accept your unchanged selfishness. What made you become more honest about accepting yourself?

N　As an actor, when reading a script and interpreting a character, it's not interesting to show only the good parts. However, when it came to my expression, I tended to exclude that perspective. But since becoming 'Non', I had to express myself as a creator. When I had to express various things without a character name, I started feeling desires like not wanting to be seen as just a 'good person' or not wanting to appear weak. From that point, instead of trying to speak well, I began to acknowledge and express my clumsiness and inability to speak perfectly, realising that this suited me better. Watching people in works, films, dramas, or variety shows and observing how they interact with others, I noticed that finding their weak points made it easier to connect with them. Therefore, I decided to look at myself more clearly and embrace those parts.

How do you plan to enjoy the next ten years of your thirties?

N　I feel more free as I get older. Although I have many immature aspects and things I can't do well, with experience, I keep discovering ways to improve. I believe there will be many more discoveries like this, and I'm excited to use all of them to enhance my expression.

のん　俳優，アーティスト　俳優を中心に音楽、映画、アートなど幅広いジャンルで活動。2017年に自ら代表をつとめるレーベル「KAIWA(RE)CORD」を発足し、音楽活動を開始。30歳を迎えた2023年、セカンドフルアルバム『PURSUE』をリリース。同年7月には「LulaBOOKS」からアートブック『Non』を発売した。

NON is active in various genres, including music, film and art, mainly as an actor. In 2017, she launched the label 'KAIWA(RE)CORD' and started her music career. In 2023, the year when she turned 30, NON released her second full-length album, *PURSUE*, and released the art book *Non* from Lula Books in July.

Memories of the Tara and agnès b.

毎に想いを乗せて

白澤貴子

フランス初の海洋に特化した公益財団法人タラ オセアンは、アニエス・トゥルブレが2003年に設立した。タラ号という海洋科学探査船は世界中を航海し、海が直面する環境問題や海洋生態系の調査を続け、2023年に20周年を迎えた。パリで2年半語学留学をしていた、エディターの白澤貴子もその活動に賛同し、これまで数々のコラボレーションを行ってきたという。日本ではまだあまり知られていないタラ号の魅力と、アニエス ベーとの思い出について。

Tara Océan, France's first public interest foundation focused on marine conservation, was established by Agnès Troublé in 2003. The Tara, an oceanographic research schooner, has been sailing around the world, investigating environmental issues and marine ecosystems. In 2023, it celebrated its 20th anniversary. Takako Shirasawa, an editor who studied abroad in Paris for two and a half years, also supports these activities and has collaborated with Tara Océan Foundation on numerous occasions. Here, we explore the allure of the Tara, which is still relatively unknown in Japan, and Takako's memories with agnès b.

夢を現実にできた
タラ オセアンとの出会い

環境問題に目覚めたのは、小学1年生の時。当時サステナビリティや環境問題という言葉はあまり浸透していませんでした。なので、興味があるということ自体変わっているという感じでした。小学生の頃から調べたり勉強したりしていて、将来は地球環境を良くする仕事に携わりたいと思っていたこともありました。結果、ファッションの世界に入り、環境問題に対しての意識や熱量は変わらずあるのに、ある視点では真逆とも言える場にいることにジレンマを感じることも多くありました。そんな時、2019年の瀬戸内国際芸術祭で開催された、タラ オセアンのプロジェクトにお声がけがありました。そこからもう4年、タラ号のオンライン船内ツアーの日仏通訳やタラ オセアン ジャパンの海洋調査体験にも参加しています。

Turning dreams into reality:
My encounter with Tara Océan Foundation

I became aware of environmental issues when I was in first grade. At that time, words like sustainability and environmental issues were not widely known. So, having an interest in such topics was considered quite unusual. Since elementary school, I have been researching and studying these issues, and there were times when I wanted to work in a field that would improve the Earth's environment. Ultimately, I entered the world of fashion. Despite my continued awareness and passion for environmental issues, I often felt a dilemma being in a field that, in some ways, seemed the complete opposite of my initial aspirations. In 2019, I was invited to participate in a project by Tara Océan during the Setouchi Triennale. For the past four years, I have been involved in various activities, such as interpreting for the online tours of the Tara schooner and participating in marine research experiences with Tara Océan Japan.

かつては冒険家が乗っていた
フランスの英雄タラ号

子どもから大人まで知らない人がいないくらい、フランスでは英雄的存在のタラ号。北極や南極、プランクトンやマイクロプラスチックなどをテーマにした探査プロジェクトや何万という生物の発見をしてきました。何が素敵って、必ずアーティストが乗船していることなんです。一緒に調査に回り、船の中で感じたことや発見したことをアートとして表現していて、その心意気にもアニエスのエスプリを感じます。彼女が本当に大切にしている船なので、知れば知るほど魅力があって。タラ号は、かつては1990年代に南極圏での調査に使っていたものなんです。そんな逸話も彼女らしくて楽しいですよね。私もいつかタ

A former vessel for explorers:
The French hero, Tara

The Tara is a heroic figure in France, known by everyone from children to adults. It has undertaken exploration projects in the Arctic and Antarctic, researching topics such as plankton and microplastics and discovering tens of thousands of organisms. What is truly wonderful is that there is always an artist on board. These artists participate in the research, expressing their experiences and discoveries through art, which reflects Agnès' spirit. The more you learn about the Tara, the more charming it becomes, as it is a ship that Agnès holds dear. The Tara was originally used for Antarctic exploration in the 90s. Agnès would tell us these stories, and she always looked very enthusiastic. It is my dream to one day board the Tara.

完璧を求めなくていい‧できることから少しずつ

タラ オセアン財団エグゼクティブディレクターのロマンが来日した際にも、マイクロプラスチックや地球環境問題に関する話を聞かせていただきました。「遅すぎることなんてない。すべて完璧にしようとしなくていい」という彼の言葉にはっとしました。マイクロプラスチックの問題も、聞けば聞くほどすごく大変で、すぐさまプラスチックをやめなくてはと感じてしまう。けれど本当にそれを実行しようと意気込んだとしても、身の回りはプラスチックに溢れていますよね。それに全部目くじらを立てるのではなくて、何か少しでも意識するってことが大事だと思うことができました。

You don't have to strive for perfection: Start with what you can do, little by little

When Romain Troublé, Executive Director of the Tara Océan Foundation, visited Japan, he also spoke to us about microplastics and global environmental issues. His words, 'It's never too late. You don't have to make everything perfect,' really struck me. The more I hear about the issue of microplastics, the more I feel that it is very difficult and that we need to stop using plastic immediately. But even if you are really determined to do it, you are surrounded by plastic. I could think that it's important to be aware of something, even if it's just a little, instead of being blind to it all.

パリの風土に馴染むアニエスベーの服

今でも多い時は年7回、少なくとも年2回はフランスを訪れています。不思議なのですが日本とフランスにいる時では着たい服が微妙に変わるんです。日本のムードのままパリに着くと、用意した服が全然気分と合わないなんていうこともあります。そんな時、パリで真っ先に駆け込むのがアニエスベーのお店。こんなものを買い足したい！と思っていくと必ず見つかるんです。生地感や色味、丈感やシルエット、どれも絶妙にこなれた抜け感があって、やっぱりフランスの空気感と合うんですよ。毎シーズン目まぐるしくスタイルが変わっていくブランドが多い中で、大きく変わることなく、でもその時々で新鮮な遊び心を感じさせる服をこれほど長くつくり続け、ずっと愛されているのはアニエスの力だなと思います。

agnès b. clothes: Blending with the Parisian atmosphere

I visit France frequently, at least twice annually, sometimes up to seven times yearly. It's strange, but the clothes I want to wear in Japan differ slightly from those I want to wear in France. There have been times when I arrived in Paris and found that the clothes I had prepared didn't match the mood at all. In such cases, the first place I rush to is an agnès b. store in Paris. Whenever I go there thinking, 'I want to add something like this to my wardrobe,' I always find exactly what I need. The fabrics, colours, lengths, and silhouettes have a perfectly relaxed, effortlessly chic feel that harmonises with the French atmosphere. In a world where many brands change their styles dramatically each season, agnès b. stands out by maintaining a consistent essence while still introducing fresh, playful elements. The enduring love for her clothes is a testament to Agnès' talent.

人として女性として目が離せないアニエス・トゥルブレ

アニエス・トゥルブレは私にとって、自由でチャーミング、そしてロックな存在。フランス人は自分に素直でいることが得意だと感じますが、彼女もまさにそう。大人になればなるほど保つことが難しくなりがちな「純粋で熱い心」を軽やかにみせるアニエスの姿にとても惹かれます。彼女の素敵な年齢の重ね方はずっと目が離せないですね。

Agnès Troublé: Unmissable presence as person and woman

Agnès Troublé is a free-spirited, charming, and rock'n'roll figure. French people are good at being true to themselves, and she embodies this perfectly. As we grow older, it becomes increasingly difficult to maintain a 'pure and passionate heart', but Agnès effortlessly showcases this quality, which I find incredibly captivating. I can't get my eyes off her wonderful way of ageing gracefully.

家族で大切にしてきたアニエスベー

アニエスベーは小さな時から着ていた馴染み深いブランド。母親が大好きで、10歳に満たない頃からずっと着ていました。子どもの頃の私は、カーディガンプレッションのイラストが描いてあるTシャツを合わせて着るのが大好きなスタイルでした。タラ号を知ってから、ブランドはもちろんのこと、人として彼女の魅力を感じるようになって、さらにアニエスベーが好きになりました。息子には赤ちゃんの頃から着せていて、彼もデニムなど何年もずっと大事に着ています。アニエスベーは親子三代で着ている大切な存在なんです。

A family treasure: agnès b.

agnès b. has been a familiar brand to me since I was young. My mother loved it, and I have been wearing it since I was 10. As a child, my favourite style was styling a T-shirt with illustrations of the snap cardigan. After learning about the Tara, I grew to appreciate the brand and Agnès as a person, which made me love agnès b. even more. I've been dressing my son in agnès b. since he was a baby, and he has treasured items like denim that he's worn for years. agnès b. is a cherished presence worn by three generations in our family.

白澤貴子　エディター　10代の頃からファッション誌の制作に携わり、多くの媒体で企画から撮影ディレクション、ライティングまでを担当。現在は雑誌にとどまらず、広告媒体のクリエイティブディレクション、ショートストーリーやコラム執筆など、アパレルのブランディング、アドバイザーなど多方面で活躍している。

Takako Shirasawa has been involved in fashion magazine production since her teenage years and has been involved in many media, from planning to photography direction and writing. Currently, Takako is active in many fields, not only in magazines, but also in creative direction for advertising media, writing short stories and columns, as well as branding and advising for apparel companies.

土地に根を張るように

岡 雄大 （株式会社Staple 代表取締役）

Taking Root in the Land
YUTA OKA (Staple Inc. CEO)

Photography: Jun Iwasaki
Text: Runa Anzai

夕刻、広島県三原港から出航する50席ほどのフェリーは、学校帰りの高校生でほぼ満席だ。黙々と読書や勉強をしたり、近所のおばあちゃんと今日の出来事を報告し合う生徒達を横目に、波の少ない穏やかな海に沈む夕日を眺める。フェリーが向かう先は、広島県尾道市瀬戸田町。瀬戸内海に浮かぶ芸予諸島の生口島と高根島というふたつの島で構成された町だ。地元の人々の生活に欠かせないフェリーが発着する瀬戸田港から島に足を踏み入れると、はじめに目に入るのが港から500メートル先の神社まで続く、しおまち商店街。この商店街の入り口にあるコーヒーショップ、OVERVIEW COFFEEをはじめ、レストランMINATOYA、日帰り銭湯を併設した宿泊施設yubuneや同じく宿泊施設のSOIL SetodaやAzumi Setodaを手掛ける岡 雄大は、5年前、ここ瀬戸田の人々と豊かな生態系に惚れ込み、事業をはじめることを決めたという。

「ホテルでもレストランでもパン屋さんでも、何かを作る時に必ずその地域に根ざしたものにしたいというのは常々考えてきました。だからこそ、観光客だけでなく、地元の人にも楽しんでもらえるような施設を作りたかった。商店街を歩いてみんなと

話していると銭湯が欲しいという人が多かったので、銭湯を作って、そうしたらお風呂上がりに美味しいビールが飲みたいというリクエストがあったので、ビールが美味しいレストランを作った。そんなふうにみんなとのコミュニケーションの中でいろいろな事業が生まれていきました」

アーキペラゴ。日本語で群島・多島海と呼ばれ、小さい島々がポツポツと点在する瀬戸内海。海外では、こうした群島に人が住むことはほとんどないのだそう。しかし、瀬戸内海ではそれぞれの島に営みや文化があり、個性も異なる。そうした地域性や人との繋がりを大切に事業を展開する岡さんの精神に共鳴したアニエスベーは「agnès b. à Setoda!」と題する企画を約1ヶ月に渡り、ここ瀬戸田で開催した。そんな岡さんがアニエスベーと出会ったのは、10代後半、アメリカに留学していた頃だった。

「もともとアニエスベーの服は好きで、学生時代にはほつれるまでバッグを使い倒した思い出があります（笑）。18歳の時、アメリカに留学したのですが、その時にサンフランシスコのユニオンスクエアにあるアニエスベーのお店に行ったんです。路

137

地裏にある、こぢんまりした店舗でその佇まいもかっこよかったですし、気さくな店員さんも素敵でした。その店員さんが、接客の際に『She likes this colour（彼女もこの色が好きなんだよね）』っていうんです。その『She』ってアニエス・トゥルブレさんのことなんですよね。こうした店員さんとブランドの創設者でありデザイナーでもある人物との近い距離感が唯一無二だと感じて、ますますアニエスベーのことが好きになりました」

学生時代にアニエスベーのお店で受けた接客。この経験が後にホテルを作った時にも活かされたと岡さんはいう。

「ホテルを作る時に究極3部屋しかなくても良いと考えています。例えば、300部屋のホテルを作っても、僕もスタッフも誰もお客さんの顔と名前が一致しないのは寂しい。部屋数が少なくても、お客さんとの距離感を大切にできる規模のホテルを作りたいと思っています。アニエスベーの店員さんがアニエスさんのことを『She』と呼ぶヒューマンスケール。そうした親密な距離感やぬくもりをホテルでも、コーヒーショップでも、レストランでも大事にしていきたいと思っています」

In the afternoon, the ferry with about 50 seats departing from Mihara Port in Hiroshima Prefecture is almost full of high school students returning from school. While the students silently read, study, or chat with local elderly women about their day, the sun is sinking into the calm, waveless sea. The ferry is headed to Setoda in Onomichi City, Hiroshima Prefecture, a town constructed on two islands, Ikuchijima and Takanejima, which float in the Geiyo Archipelago of the Seto Inland Sea. Stepping off the ferry—a vital transportation for locals—at Setoda Port, the first thing that catches the eye is Shiomachi Shopping Street, which stretches from the port to a shrine 500 meters away. Captivated by the people of Setoda and its rich ecosystem, Yuta Oka decided to start his ventures here five years ago, including the coffee shop OVERVIEW COFFEE at the entrance of the shopping street, the restaurant MINATOYA, the accommodation facility 'yubune' with a day-use bathhouse, and other lodging options like SOIL Setoda and Azumi Setoda.

'Whether it's a hotel, restaurant, or bakery, I've always wanted to create something rooted in the local community. That's why I aimed to build facilities that both tourists and locals can enjoy. While walking through the shopping street and talking to everyone, I found that many people wanted a public bathhouse, so I built one. Then, there were requests for a place to enjoy a delicious beer after a bath, so I created a restaurant that serves great beer. In this way, various ventures emerged through communication with the community.'

Archipelago—in Japanese, it's called *guntō* or *tatōkai*, referring to the scattered small islands in the Seto Inland Sea. Outside of Japan, it's rare for people to inhabit such archipelagos. However, each island has its own life, culture, and unique characteristics in the Seto Inland Sea. Resonating with Yuta's spirit of valuing local identity and connections with people, agnès b. held a project named 'agnès b. à Setoda!' for a month here in Setoda. Yuta first encountered agnès b. in his late teens when he studied abroad in the United States.

'I've always loved agnès b.'s clothes, and I remember using their bags until they were frayed during my student days. (Laughs) When I was 18, I studied abroad in the United States and visited the agnès b. store in Union Square, San Francisco. It was a small shop tucked away in an alley, and its appearance was really cool. The friendly store staff were also wonderful. During our interaction, one of the staff members said, "She likes this colour", referring to Agnès Troublé. The close connection between the staff and the brand's founder and designer felt unique, and it made me love agnès b. even more.' This customer service Yuta encountered at the agnès b. store during his student days influenced him when he opened his hotels later.

'When creating a hotel, I believe it's fine to have only three rooms if that's what is needed. For example, if I built a hotel with 300 rooms, it would be sad if neither the staff nor I could recognise the faces and names of the guests. I want to create a hotel that, even with fewer rooms, maintains a close connection with its guests. The way the agnès b. store staff referred to Agnès as "she" embodies a human scale. I want to cherish that intimate distance and warmth in my hotels, as well as in my coffee shops and restaurants.'

岡 雄大　株式会社Staple 代表取締役　瀬戸田と日本橋に拠点を置きまちづくりを行うソフトデベロッパー、株式会社Staple代表取締役。2018年の設立以来、日本橋兜町・K5のプロデュースやSOIL Setoda、SOIL Nihonbashiなどの企画・開発・運営を行う。また共同代表を務める別会社を通してAzumi Setoda、yubuneなどの宿泊施設も手掛ける。

Yuta Oka is the CEO of Staple Inc., a 'soft developer' company based in Setoda and Nihonbashi that engages in town planning. Since its establishment in 2018, the company has been involved in the planning, developing, and operating projects such as K5 in Nihonbashi Kabutocho, SOIL Setoda, and SOIL Nihonbashi. Through another company where he serves as co-president, Yuta also manages accommodations like 'Azumi Setoda' and 'yubune'.

永年愛せるベーシックなアイテムを基盤に、チャーミングなアイデアで支持を集めるスタイリストの金子綾。アニエスベーの代名詞ともいえるカーディガンプレッションを使った3つのスタイリングを提案。コーディネートの考え方から服選びのポイントまで、彼女のオリジナリティに迫った。

Stylist Aya Kaneko, known for her charming ideas based on timeless, beloved essential items, presents three styling suggestions using agnès b.'s iconic snap cardigan. From her approach to coordination to her tips on choosing clothes, we delve into Aya's original perspective.

Forever Chic and Charming

シックに宿る愛らしさ

金子 綾

娘とシェアするカーディガンプレッション
アニエスベーのカーディガンプレッションは、旅行の必需品です。コットンなので洗えるし、夏でもクーラーが寒かったりするので重宝します。左腕を通さずアシンメトリーになるように着てみたり、ビスチェのように着てみたり、自由自在にアレンジが利くのが嬉しい。最近カーディガンを買い替えたのですが、前に着ていたものは小学生の娘が袖を折って着ています。カーディガンプレッションはもちろん、傘もアニエスベーのものを持たせていて、要所要所でブレないようにしています（笑）。

Sharing the snap cardigan with my daughter
The agnès b. snap cardigan is an essential item for travel. Since it's made of cotton, it's washable and perfect for cold, air-conditioned places in summer. I love its versatility; you can wear it asymmetrically by not putting your left arm through or even wear it like a bustier. Recently, I bought a new one, and the one I used to wear is now worn by my elementary school daughter, who folds up the sleeves. Along with the snap cardigan, I also give her an agnès b. umbrella, making sure she stays stylish all the time. (Laughs)

これからもずっと、私のそばにある服

5パターンくらい着回しがイメージできたとき、服を買うことが多いです。10年後の自分を想像してみて、これを着ていたら可愛い、素敵だなと思えるかどうかが大事。黒の服を好んで着るのですが、アニエスベーの黒は、カジュアルすぎずシックすぎない、かといって地味でもない、ちょうどいいバランス。ただ、年齢やキャリアを重ねるにつれて、黒だけだと迫力がでてしまうこともあるので、最近はチャーミングな印象になるようにリボンやパールをアクセントに取り入れるようにしています。

Clothes that will always be by my side

I often buy clothes only when I can envision about five different ways to style them. I need to imagine myself 10 years from now and think, 'Would this look cute and stylish on me?' I prefer wearing black clothes, and agnès b.'s black is perfect—it's not too casual, not too chic, and definitely not boring; it has just the right balance. However, as I age and progress in my career, wearing only black can sometimes feel overwhelming. So, recently, I've been incorporating accents like ribbons and pearls to create a more charming impression.

アイデアのまま、自由に服を着ること

スタイリングの醍醐味は、ベーシックなアイテムにどう捻りをつけるか。普通のカーディガンをビスチェのように巻くのは意味がない。アニエスベーのカーディガンプレッションだからこそ、ボタンがステンレスのように光って目を引くし、ボタンとボタンの間の距離が近いので、アレンジ次第でちゃんとポイントになってくれるんです。要素が多いけれど品がよくて、どんなスタイルにも寄り添ってくれるので一年中手放せません。

To dress freely, just as ideas come to mind

The essence of styling lies in adding a twist to essential items. Wrapping an ordinary cardigan like a bustier doesn't make sense. It's because it's an agnès b. snap cardigan that the buttons shine like stainless steel and catch the eye, and the short distance between the buttons allows for versatile styling. Despite having many elements, it maintains a sense of elegance and complements any style, making it indispensable all year round.

定番には、品よく二面性を

シックなものに、シックなものは合わせない。それが私のスタイルです。このワンピースは綺麗なので、あえてスポーティな要素のあるサンダルを合わせてカジュアルダウンしました。ギャップが大事ですよね。タイトスカートも私の定番アイテムのひとつなのですが、スウェットを合わせたりキャップを被ったりして、その日の気分でいかにコンサバを崩せるか考えています。

Two ways of wearing classic items

I don't pair chic items with other chic items. That's my style. This dress is quite elegant, so I deliberately paired it with sporty sandals to dress it down. The contrast is important. Tight skirts are also one of my go-to items, but I like to mix them up with a sweatshirt or a cap and think about how I can break the conservative look based on my mood for the day.

パーソナリティは小物に委ねて

自分の印象をカジュアルに見せたいわけでも、奇抜に見せたいわけでもないので、小物に関してもシックで上品な印象になるものを選びます。あと、フレンチネイルを20年以上続けているのもこだわりです。カジュアルな格好をしたときも、品よくまとまるかなと思って。人見知りなので怖そうと言われることもあるのですが（笑）だからこそ迫力がですぎないように、ささいな印象にも気を配っています。

Personalise an outfit with accessories

I don't aim to appear casual or extravagant, so I choose accessories that create a chic and elegant impression. Also, I've been doing French nails for over 20 years. Even when dressed casually, they help convey a pronounced sense of elegance. Since I'm shy and sometimes appear intimidating, I pay attention to the subtle details to avoid appearing too intense. (Laughs)

変化を許容してくれる、本質的なデザイン

アニエスベーのお洋服って、気分ではないときがあっても、また手に取るタイミングが来るんですよね。そのときには着方が変わっていることもあります。たとえばカーディガンプレッションは、全部ボタンを留めてしまえばインもできるし、ニットとしても着ることができる。そうやってアレンジを変えて着るのも楽しいですよね。普遍的で飽きがこないデザインならではだと思います。

Accepting change with essential design

With agnès b. clothing, even if there are times when it doesn't suit your mood, you always come back to it. And when you do, the way you wear it might have changed. For example, with the snap cardigan, you can button it up entirely and wear it tucked in or as a knit. It's fun to change the way you style it. I think it's because of the timeless design that is never boring.

等身大のわたしを受け入れていく

歳を重ねると、カジュアルなスタイルはくたびれて見えるときがあるので難しい。白Tにデニムの場合は、Tシャツに少し張りがあったり、カーディガンプレッションのように要素があるものを合わせるとちょうどいいバランスになります。もう10年、20年経って、シワができてきたりヘアがグレーになったりしたら、それもアクセントとして活かしながら、海外のおばさまたちのように自由なコーディネートも楽しんでみたいなと思います。

Embracing the Real Me

As I age, casual styles can sometimes look tired, making them more challenging to wear. For example, when wearing a white T-shirt and jeans, it is nice to have a T-shirt with a bit of structure or pair it with something detailed like a snap cardigan for the perfect balance. In another 10 or 20 years, when I get wrinkles, and my hair turns grey, I want to embrace those as accents and enjoy creating free-spirited outfits like the stylish grannies in other countries.

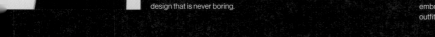

金子 綾　スタイリスト　1979年生まれ。静岡県浜松市出身。長年第一線で活躍する人気スタイリスト。『VERY』や『Oggi』といったファッション誌をはじめ、カタログや広告、ブランドとのコラボレーション＆ディレクションなど、その活動の幅は多岐にわたる。

Born in 1979 in Hamamatsu, Shizuoka Prefecture, Aya Kaneko is a popular stylist who has been active on the front lines for many years. Her work spans various activities, including fashion

RINKO
KAWAUCHI

Seasonal Change
of Clothing

衣替えのたびに
— 川内倫子

Photographer

Interview

Interview

Interview

2019年にアニエスベーのカーディガンプレッション誕生40周年を記念して、「自由にカーディガンを撮影してください」という依頼をいただいたとき、大学生だった頃の教室の風景が脳裏に浮かんだ。そのカーディガンが大流行して、流行に敏感な同級生の多くがさまざまな色のそれを着ていたからだ。自分は貧乏学生だったので手が届かず、ただ羨ましく見ていただけだったが。

どのように撮影するのか悩んだが、自分の身近な人に着てもらうのがしっくりくるような気がして、まずは娘に着せて撮影することにした。そのうちに物足りない気がして知人の7歳の娘さんの薫ちゃんにモデルになってもらった。カーディガンが誕生した40年前は自分が7歳だったので、私的な物語として当時の自分を薫ちゃんに重ねてみたのだ。撮影中に薫ちゃんの後ろ姿を見ていると、背格好が似ていることもあり、時間を飛び越えてあの頃の自分を眺めているような錯覚があった。小学校に通い始めたばかりで学校に馴染めず、息苦しさを感じていた自分が目の前を走っているようだった。そして娘と薫ちゃんが自分の憧れであったスナップカーディガンを着ている姿は、自分の記憶のなかの幼い自分が癒されていくような感覚もあった。

あれから数年後、撮影で娘が着用したカーディガンは、成長した娘には小さくなったので、友人の娘さんに譲ることにした。思い出も一緒になくなってしまいそうな気がして、送るための段ボール箱に一度入れてから逡巡し、やはり手元に置いておこうかといったん取り出したが、いやしかし着てもらったほうが服も活かされるだろう、このカーディガンは幅広くさまざまなタイプの人に似合いそうだし、と思い

直して箱に戻した。そうやって小さくなった服は毎年増えていき、その時々に気に入って着せていた服は早くて半年、遅くても2年で着られなくなる。それらは衣替えの時期になると、娘よりも小さなお子さんがいる友人知人宅へまとめて送る。その度に、もう戻ってこない日々を振り返り、無事にここまで育った喜びも同時に噛み締めることになる。

時々着ていた写真を見返しては、もうこの頃のあの子には会えないのだと寂しいような気もするが、目の前にいる成長した娘の姿を見ると、すぐに現実に引き戻される。いまちょうど7歳になった娘は、7歳だった頃の自分には似ていない。背格好もそうだが、学校へ行くのを楽しんでいるし、友達をつくるのも得意だ。4歳頃からは服も自分でコーディネートするようになり、時折こちらが他のものを薦めても、気分が沿わないと頑なに拒否される。自分が気に入って買った服も、結局一度も着てくれずにサイズオーバーしたことが何度もある。でも自分の好きなものに対して意志を通すということはこれから先の人生で必要な要素でもあるだろう。つい自分が同じ歳だった頃と比較して世話を焼き過ぎそうになったりもするが、彼女は彼女らしく生きていくべきだし、サポートに徹するべきだと自分を諫めるときもある。

そのうちに自分の身長も追い越すだろう。いまある自分のクローゼットのなかの服をシェアできる日もすぐだろう。でもその時間もきっと短くて、彼女とこの家に一緒に住む時間も長くないのだろう。その日のことを考えると今からすでに寂しいが、それよりも彼女と一緒に生活し、成長していく姿をただ眺めることができる、今現在を享受したい。

In 2019, when I was asked to freely photograph the agnès b. snap cardigan to celebrate its 40th anniversary, the image of my college classroom came to mind. The cardigan was incredibly popular at that time, and many of my fashion-conscious classmates wore it in various colours. Being a poor student, I couldn't afford one myself and could only look on with envy.

I struggled with how to shoot it, but it felt right to have someone close to me wear the cardigan, so I decided to start by photographing my daughter in it. Eventually, I felt that it wasn't enough and I asked my friend's seven-year-old daughter, Kaoru, to be my model. 40 years ago, when the cardigan was released, I was seven years old, so I intertwined my personal story by overlaying my younger self with Kaoru. Watching Kaoru's back during the shoot, I had a moment of déjà vu as her stature resembled mine, creating an illusion of looking at my younger self across time. It felt as if the me who had just started elementary school, struggling to fit in and feeling suffocated, was running right in front of me. Seeing my daughter and Kaoru wearing the snap cardigan I had once longed for was a healing experience, as if my childhood self within my memories was being comforted.

A few years later, the cardigan my daughter wore for the shoot had become too small for her as she grew, so I decided to pass it on to another friend's daughter. I felt like the memories associated with it might disappear too, and after putting it in a cardboard box to send it off, I hesitated and thought about keeping it. However, I reconsidered, thinking it would be better if someone else could wear it and bring it to life. Thinking that the cardigan seems to suit a wide variety of people, I put it

back in the box. Every year, the number of outgrown clothes increases, and my daughter grows out of the outfits I carefully choose in as little as six months to at most two years. During the seasonal wardrobe change, I pack up these clothes and send them to friends with younger children. Each time, I reflect on the days that will never return and simultaneously savour the joy of seeing my daughter grow up safely up to this point.

Sometimes, when I look back at photos of her wearing those clothes, I feel a twinge of sadness, realising that I'll never see her as she was then again. But seeing my grown-up daughter right in front of me quickly brings me back to reality. My daughter, who has just turned seven, is not like how I was at her age. Not only is her build different, but she also enjoys going to school and is good at making friends. Since she was about four, she has been coordinating her own outfits, and occasionally, she stubbornly refuses my suggestions if they don't match her mood. There have been many times when she outgrew clothes I bought for her without ever wearing them. But having a strong will about her preferences is an important trait for her future. I often find myself comparing her to how I was at her age and becoming overly protective, but I remind myself that she should live her life in her own way and that I should focus on supporting her.

In time, she will probably outgrow me in height. The day when we can share the clothes in my closet will come soon, but that period will likely be brief. The time we will live together in this house won't be long either. Thinking about that day already makes me feel lonely, but more than that, I want to cherish the present moment, living together and watching her grow.

川内倫子 写真家 1972年、滋賀県生まれ。2002年に『うたたね』『花火』で第27回木村伊兵衛写真賞受賞。2023年にソニーワールドフォトグラフィーアワードのOutstanding Contribution to Photography（特別功労賞）を受賞するなど、国際的にも高い評価を受け、国内外で数多くの展覧会を行う。主な著作に『Illuminance』（2011年）、『あめつち』（2013年）、『Halo』（2017年）など。近刊に写真集『やまなみ』『橙が実るまで』（田尻久子との共著）がある。2022～2023年に東京オペラシティアートギャラリーと滋賀県立美術館で個展「川内倫子：M/E 球体の上 無限の連なり」を開催した。

Born in Shiga Prefecture in 1972, Rinko Kawauchi won the 27th Ihei Kimura Photography Award in 2002 for her works *Utatane* and *Hanabi* and was awarded the Outstanding Contribution to Photography at the Sony World Photography Awards in 2023. Rinko has held numerous exhibitions in Japan and abroad. Her major publications include *Illuminance* (2011), *Ametsuchi* (2013), and *Halo* (2017). Her recent publications include the photo-books *Yamanami* and *Daidai-ga-minoru-made* (co-authored with Hisako Tajiri), and her solo exhibition *Rinko Kawauchi: M/E On this sphere, Endlessly interlinking* was held at Tokyo Opera City Art Gallery and the Shiga Prefectural Museum of Art from 2022 to 2023.

ニューヨーク出身のリン・スレーターは、大学教授としての長いキャリアを終え、61歳でファッションについてのブログを書きはじめた。"Accidental Icon" という名前のブログでリンは、度々アニエスベーのことを紹介していたのだ。それらの記事をアニエス・トゥルブレ本人が読んだことをきっかけにふたりは親交を深めることになった。世界中にファンを持つファッションアイコン、そして作家であるリンが、お気に入りのカーディガンプレッションとアニエス・トゥルブレとの思い出を綴る。

Staying Curious

好奇心を持ち続けること
リン・スレーター

　私がアニエスベーを知ったのは、大学教授だった頃、新学期を迎えるための服を探していた時でした。ウェイターの制服を彷彿とさせるようなアニエスベーのヴィンテージの黒いジャケットを見つけたんです。制服は、想像力次第で個性的な着こなしができるでしょう? アニエスベーの服は日常のありふれたものを尊重しつつ、それをちょっと特別なものに昇華させてくれるから好きなんです。

　アニエスベーのカーディガンプレッションは特にお気に入り。カッティングや生地感も素晴らしいんです。スウェットシャツのような気軽な感じもありながら、おしゃれをしてお出かけしたい時にもフィットする。異なる素材やシルエットがあるおかげで、どんな人にも似合うし、決して古びかない。時代の移り変わりと共に着る人に寄り添ってくれる服なんです。

　人生の節目にカーディガンプレッションを購入しています。最初に購入したのは、40歳の誕生日を迎えた時の自分へのプレゼント。ややゆったりとした作りのブラックのもの。そこから色々なシルエットを購入していて、レザーのものや最近はこのコバルトブルーのものを購入しました。歳を重ねるにつれてブラックよりもブルーが気になってきて、よく着

るようになりました。

　お気に入りの服をデザインしてくれたデザイナーに私は日々魅了され、彼らについてできる限り理解を深めたいと思っています。アニエス・トゥルブレについて調べていると、ファッションデザイナーだけでなく、映画監督やプロデューサー、さらにはアーティストの支援や環境保護活動家というさまざまな側面を持っていることに気がつきました。時流や周りの環境に常に好奇心がある彼女がどこか私と似ていると感じることもあります。出会うべくして出会ったのだと思います。

　現在ブログはお休みしていて、ソーシャルメディアやファッションの世界とは少しだけ距離をおいています。都会から田舎に引っ越してきて、場所の変化が私の着こなしや行動、欲求にどういった影響を与えるのか、これからが楽しみです。田舎は都会とまったく異なるエネルギーがあって、生活もガラリと変わりました。年齢のせいもあるかもしれないけれど、とてもゆっくりと仕事をするようになりました。どんな環境で暮らすか、仕事をするかは、私たちの人生を形成する重要な要素です。そして、常に好奇心を持ち、新しい道筋を示してくれるようなものにオープンであることを心がけています。

After a long career as a college professor, at the age of 61, New Yorker Lyn Slater started a blog to write and think about fashion, under the name of 'Accidental Icon'. On her blog, Lyn often featured agnès b. in her articles that caught the attention of Agnès Troublé herself, and the two were eventually introduced. A fashion icon and writer with fans worldwide, Lyn shares her memories with agnès b. and why the snap cardigan is her favourite item.

I first became aware of agnès b. while shopping for work clothes. I was styling my look for a new semester as a professor. I found a vintage agnès b. black jacket that put me in mind of a waiter. I am a fan of uniforms because if you have imagination they can be worn in unique and individual ways. agnès b. clothing honours the everyday and ordinary yet makes it something just a bit more.

The snap cardigan does just that: the cut and texture, especially the fleece, puts you in mind of a favourite, comfortable old sweatshirt that you usually only wear at home. It takes the comfort of the favoured piece and elevates it so you can wear it outside and wherever else you need to go. The different materials and silhouettes allow you to make different choices depending on what you want to express in any particular moment of your life. It's universal and ageless that way. It's a garment that can change as you do: it lives with you, no matter how you change over time.

It seems I've acquired a snap cardigan during each phase of my life. The first one I got was a gift when I was 40. It was black fleece with a wider cut. From then I started to collect them in their various silhouettes. I even have one in black leather and I recently acquired the bright cobalt blue one. It is meaningful to me as in my older life I seem to prefer the colour blue over black.

I am fascinated by the person who designs a garment I love and try to find out everything about them. While searching about Agnès Troublé, I discovered a woman, who is not only a fashion designer, but also a filmmaker and producer, a supporter of artists and an environmentalist. Like me, she is always curious about her times and surroundings, and this curiosity naturally encourages self-reinvention.

Right now, I am taking a break from my blog and not really engaging with social media or the fashion world. I have moved out of the city and am living in the country, and I am observing how this change of location is impacting what I wear, what I am doing and what I want to do. It is a very different energy, so my work is slowing down, partly because I am older, also because of where and how I live now. Our surroundings shape us, and curiosity is essential to stay relevant. I try to remain curious and open to the things that may present themselves as providing me with perhaps a new map to follow.

リン・スレーター　作家　大学教授として教鞭を執った後、2014年に「Accidental Icon」というブログを開設。以降、モデルや作家、コンテンツクリエイターとして活動する。2024年3月に著書『How To Be Old』を刊行。

After teaching as a university professor, Lyn Slater founded the blog 'Accidental Icon' in 2014. Lyn has since worked as a model, author, and content creator, and her new book, *How To Be Old*, was published in March 2024.

いくつになっても自分の「好き」に正直でいることは、シンプルだけれど難しい。
ミュージシャン、シンガーソングライターのカネコアヤノが大切にしている
自分を見失わないために譲れないこと。

嫌いだから、好きがある

カネコアヤノ（ミュージシャン, シンガーソングライター）

The Courage to Be Honest
AYANO KANEKO (Musician, Singer-songwriter)

Photography: Reiko Toyama
Hair & Make-up: Marico Takaki
Text: Rei Sakai

アニエスベーと出会ってからもう長いそうですね。

K　気づいたら知っていたくらい、昔から馴染みのあるブラ
　　ンドです。今日着たブラックのカーディガンとワンピース
　　は、お母さんが誕生日に買ってくれたもの（P154,P155,
　　P157）。カーディガンはずっと憧れていたもので、私に
　　とって初めてのアニエスベーかな。ちなみにお母さんと
　　お揃いなんです。

もう一着はワンピースですね。私服で40着以上のワンピース
をお持ちだと聞きました。

K　このワンピースはポケットとか、爪みたいな柄が可愛く
　　て。いつでも着やすいし動きやすくて気に入っています。
　　私服のワンピースは、最近断捨離しました。でも思い出
　　があったり、買った時の記憶があるものは残っています
　　ね。あとは洋服の表情というか、個性が見えるものはや
　　っぱり手放せません。たとえ着ない年があっても、次の
　　年に急に「いまいいかも！」っていう気分になることが多
　　くて。やっぱりなんだかんだ気に入っているんだなって
　　思います。

着ている服のシルエットに自信をもらえるような、精神的な影
響ってありますか？

K　めちゃくちゃありますね。テンションがすごく落ちている
　　日とかも、好きな服を着たり、好きなディテールのもの
　　に包まれていると大丈夫な気がする。戦闘服みたいな
　　感じなのかも。パワーになります。

今日着たカーディガンやワンピースにも、そういったパワーを
感じますか？

K　あります！カーディガンを着ると、よし！ってちょっと背
　　筋が伸びる感じがしますね。これ着ている私、って感じ。
　　結局、そのメンタルって大事だなってすごく思います。ア
　　ニエスベーが好きな理由はそこですね。アニエスベーを
　　着ている人ってみんなパリッとしているというか。そうい
　　うところに惹かれているのかもしれません。

ファッションと音楽。どちらも自分の中から出てくる表現です
が、両方に共通した自分らしさを挙げるとしたらなんでしょう。

K　どちらも、街を歩いていて吸収していくものではあるな

と思います。洋服も、街を歩いている人とかを見て可愛いなとか、こんな合わせ方もあるのかって思うし、歌は散歩している時とか、友達と会って話をしていると生まれてきます。

普段の日常の中で感じたこと、見つけたものを大切にしている。

K　そうですね。自分じゃないことというか、演出するのが苦手なんです。だから、ステージで着たい服も日常で着たい服と同じ。私、嫌いなこととかやりたくないことがすごくはっきりしていて。音楽をやっている時も生活している時も、嫌なことが見えないと、好きなこととかやりたいことが見えてこない。常に嫌だなって思うことにちゃんと向き合うようにしていますね。だから、私はこれがやりたかったのかって。

嫌なことをひとつ挙げるとしたらどんなことですか？

K　中途半端なこと。あと、自分の意見がなくなっちゃって、自分がいなくなっちゃうんじゃないだろうかみたいな瞬間。そこで飲み込まれないようにしないといけない。やっぱり「はい」と言うことよりも「嫌です」って言う方が私は大変なタイプだから、そこでどれだけちゃんと言えるか、みたいな。

社会で生活していると、本当にたくさんの意見が入ってきて、

その中でなるべく自分に素直にいるというのは、すごく難しいことですよね。

K　そうなんです、全然できない瞬間もあります。でも、できるだけやっぱり言いたい。言いたいというか、その一回の嫌なことで、「はい」って言ってしまうことで、今まで自分が積み上げてきたことがブレたり、何か崩れるちょっとの溝とか傷になるのがすごく嫌で。だから、そこはすごく戦いたいところですね。

守りたいですね。

K　はい。私、ピンクがすごく好きなんですけど、何歳になってもピンクや赤、真っ白なワンピースを着ていられる自分でいようっていうのは思っていて。それこそ今日のカーディガンとかタイツもそう。それをちゃんと、ずっと好きでいられるように。頑張ることではないけど、好きに素直でいようとは思っています。

洋服に限らず、タイムレスに愛されるものには何があると思いますか？

K　創作者の決意じゃないですかね。その覚悟。長く残っているブランドもそうだし、長く愛されている音楽も、結局そこがすごく大きい気がします。

No matter how old you get, staying true to what you 'love' is simple yet difficult. This is what singer-songwriter Ayano Kaneko values to ensure she doesn't compromise and lose herself.

It is said that you've known agnès b. for a long time now.

K   Yes, it's a brand I've known for as long as I can remember. The black cardigan and dress I'm wearing today were birthday gifts from my mother (P154, P155, P157). I had always admired the cardigan, and it was my first agnès b. piece. My mother has the same one, too.

And the other piece is a one-piece dress. Is it true that you have over 40 dresses in your personal wardrobe?

K   Yes, this dress has cute details like pockets and a claw-like pattern. It's comfortable and easy to move in, which is why I like it so much. Recently, I did some decluttering of my dresses. However, I kept the ones that hold memories or have sentimental value from when I bought them. Also, I can't let go of clothes that have a unique character or personality. Even if there's a year when I don't wear them, I often find myself suddenly thinking, 'This feels right now!' the following year. It made me realise how much I actually liked them.

Does the size of the silhouette of your clothes affect you mentally, like giving you confidence?

K   Absolutely. Even when I'm feeling down, wearing clothes I love or being wrapped in details I like makes me feel like everything will be okay. It's almost like an armour. It gives me power.

Do you feel that kind of power from the cardigan and dress you're wearing today?

K   Yes, I do! When I wear the cardigan, I feel more upright and confident. It's like, 'This is me, wearing this cardigan.' I realise how important that mentality is. That's one of the reasons I love agnès b. People who wear agnès b. seem to have a certain sharpness to them. I think that's what draws me to it.

Fashion and music are both forms of expression that come from within. What is a common thread they share that is unique to you?

K   My fashion style and music are influenced by what I absorb and take in while walking around the city. With clothes, I notice cute styles or new styling combinations from people I see on the streets. Songs come to me when I'm out for a walk or having conversations with friends.

You value what you feel and discover in your everyday life.

K   That's right. I'm not good at pretending to be somebody I'm not. That's why the clothes I want to wear on stage are the same as the ones I want to wear in everyday life. I have a very clear sense of what I dislike and don't want to do. When making music or just living everyday life, if I don't acknowledge what I don't like, I can't see what I do like or want to do. I always try to face head-on the things I find unpleasant. I think to myself, 'Is this really what I wanted to do?'

If you had to name one thing you dislike, what would it be?

K   Leaving things unfinished or being half-baked. Also, there are moments when I feel like I've lost my own opinion and, consequently, a sense of myself. I have to make sure I don't get swallowed up in moments like that. It's more challenging for me to say 'no' than to say 'yes', so it's about how well I can express my true feelings in those situations.

In our society, we're constantly exposed to a multitude of opinions, and staying true to yourself is incredibly difficult, isn't it?

K   Yes, there are definitely moments when I can't manage it at all. But I still want to speak up as much as possible. It's not just about wanting to say something; by saying 'yes' to even one unpleasant thing, everything I've built up can become shaky or develop small cracks or wounds. I really don't like that. So, that's something I want to fight for.

It's something we want to protect.

K   Yes. I love pink, and I want to be the person who can wear pink, red, or a white dress no matter how old I get, like today's cardigan and tights. It's not something I have to work hard for; I just want to stay true to what I love.

Not only for clothes, what do you think makes something timelessly loved?

K   It's the same for brands and music that have been loved for years. Ultimately, I think it's the determinator of the creator.

カネコアヤノ　ミュージシャン, シンガーソングライター　弾き語りとバンド形態でライブ活動を行っている。2021年に発表したアルバム『よすが』は第14回CDショップ大賞2021に入賞。同年に初の武道館ワンマンショーを開催。2023年1月にフルアルバム『タオルケットは穏やかな』、3月に弾き語りアルバム『タオルケットは穏やかな ひとりでに』を発売。同年1月に日本武道館ワンマンショー2days、3月に初の大阪城ホールワンマンショーを開催し、7月に「カネコアヤノ Hall Tour 2023 "タオルケットは穏やかな"」を成功させた。

Ayano Kaneko is a singer-songwriter who performs live, both solo and with a band. Her 2021 album *Yosuga* was awarded the 14th CD Shop Award 2021. In the same year, Ayano held her first solo show at the Nippon Budokan. In January 2023, Ayano released her latest full album, *A Towel Blanket is Peaceful* and in March, her solo album, *A Towel Blanket is Peaceful - Solo*. That year, Ayano performed a two-day solo show at the Nippon Budokan in January, held her first solo show at Osaka-jo Hall in March, and completed the Ayano Kaneko Hall Tour 2023 in July.

157

Photography: Naoki Usuda
Hair & Make-up: Ken Nagasaka
Text: Rei Sakai

MEIRIN
Thinking While Making

ソロプロジェクト「ZOMBIE-CHANG」として
音楽活動を行う傍ら、モデルやアーティストとして
活動するメイリン。フランスのミュージシャンとの
コラボレートや、創作活動のひとつである
編み物のこと、それぞれに共通する
ものづくりの向き合い方について。

パリを拠点に活動するAgar Agarとコラボレーションされていましたね。

M　今回はコラボレートで1曲を作ったのではなく、お互いにトラック
　　を送り合って歌詞をつけて戻して、ZOMBIE-CHANGから1曲、
　　Agar Agarから1曲出すという形にしました。タイトルの「T'inquiète
　　pas」は、日本語で「大丈夫、心配しないで」という意味。その時
　　覚えたばかりのフランス語を使いたくてこのタイトルにしました。
　　歌詞のテーマを伝えて、じゃあこういう音も入れようかと相談しな
　　がら、シンセサイザーを足してもらったりして進んでいきましたね。

過去のコラボレート作品もみなさんフランスの方ですね。言語や文化
の違いがあるなかで、第三者と制作する難しさはありますか?

M　言語はあまり関係なくて、制作面でのエゴイズムを共有すること
　　が難しいなと思いました。お互いのエゴとエゴをどう馴染ませる
　　かが大事。今回のAgar Agarのように、トラックを交換し合って、
　　それぞれ1曲ずつ発表する形はいいなと思いました。トラックを
　　渡したら、後はもう好きにやってみてって。

誰かと一緒に作ることで気がついたことや、発見はありましたか。

M　やっぱり一人って最強だなと。より一層、一人でやろうという気

While pursuing her music career as the solo project ZOMBIE-CHANG, Meirin also works as a model and artist. Meirin collaborates with French musicians and engages in creative activities such as knitting. Here, we explore Meirin's approach to craftsmanship and the commonalities in her creative processes.

持ちが強くなりました。自由を求めるエナジーになるし、自分の
エゴを再認識するきっかけにもなります。でも、今まで私のこと
を知らなかった人に届けられるという意味では、やっぱりコラボ
レートは嬉しいです。フランスのアーティストと作ったら、フラン
スの人が私のことを知るきっかけになる。それは向こうにとっても
同じで、お互いにとって世界を広げるきっかけになるのはいいこ
とだなと思います。

編み物もされていますよね。創作のアイデアは、どこから来るのでしょ
うか。

M　人の手で作られて販売されているものは、時間がかかる分お値
段もするじゃないですか。でも自分でも作れるんじゃないかと思
ったのがきっかけで編み物をはじめました。元々何かを作ること
がすごく好きなので、自分の好きなように考えながらものづくりを
するのが楽しい。作曲と同じなんですよね。さっき、Agar Agar
の曲のタイトルは覚えたてのフランス語を使ったと話したのです
が、編み物も覚えたての技術を使ってどんなものができるか試し
ている感覚です。

技術を学びたいなと思った時、応用が効きそうなものを選んだりしてい
ますか?

M　そうですね。この前はニットでプリーツスカートを作りたくて、じ
ゃあ今持っている技術でどうやったら作れるかなと考えて。この
編み方とこの編み方を組み合わせたらプリーツのデコボコになり
そうとか、サイズ感が難しそうだから巻きスカートにしようとか。
いつも作りながら考えていますね。音楽も一緒です。

ゴールを決めないようにしている。

M　はい。間違えても修正すればいいだけなので、後で何とかしよう
って。ゴールを決めない方が、気が楽で続けられるんです。多分、
編み物を挫折する人は、編み図で正解を見てから始めて、やっ
てみたらその通りにできないからやめてしまうんじゃないかなと思
っていて。曲も、最初から完成形が見えてることってないんです
よね。完成を目指すのではなくて、作りながら考え続けて、行
き着いた先が一つの完成なのかなと思っています。

Your work includes a collaboration with Agar Agar, a band based in Paris.

M　This time, instead of creating one song together, we decided to send tracks back and forth to each other, add lyrics, and then each release a song—one from ZOMBIE-CHANG and one from Agar Agar. The title 'T'inquiète pas' means 'It's okay, don't worry'. I had just learned this phrase in French, and I wanted to use it. We discussed the theme of the lyrics and added elements like synthesisers as we went along, incorporating suggestions and ideas from both sides.

Your past collaborations have all been with French artists. Is it challenging to create with others?

M　Language isn't much of an issue; I found it more challenging to share the egotism that comes with the creative process. It's important to figure out how to blend each other's egos. I thought our approach with Agar Agar, where we exchanged tracks and each released a song, worked well. Once I handed over the track, I let them do whatever they wanted.

Did you notice or discover something new by creating with someone else?

M　I realised that being solo is truly the best. This realisation strengthened my desire to work independently even more. Collaborating fuels my energy for freedom and allows me to reaffirm my ego. However, collaborations are exciting in terms of reaching people who didn't know about me. Working with French artists, for example, introduces me to the French audience and vice versa. It's a great way for both parties to expand their worlds.

You also knit, don't you? Where do your creative ideas come from?

M　Handmade items sold in stores can be expensive because they take time to make. I started knitting because I thought I could make these myself. I've always loved creating things, so it's fun to make things while thinking freely about what I like. It's similar to composing music. Earlier, I mentioned that I used a newly learned French phrase for the title of the song with Agar Agar. Knitting is the same for me—I want to experiment with new techniques.

When you want to learn a new technique, do you choose ones that seem adaptable?

M　Yes, exactly. Recently, I wanted to make a pleated skirt by knitting, so I thought about how I could create it using techniques I already know. I considered combining different knitting patterns to achieve the pleated texture, and since the sizing seemed tricky, I decided to make it a wrap skirt. I always think things through as I create, and it's the same with music.

And you avoid setting a specific goal.

M　Yes. If I make a mistake, I can just fix it later, so I think, 'I'll figure it out somehow.' Not setting a specific goal makes it easier and more enjoyable to keep going. I think people who give up on knitting often start by looking at the pattern and aim for perfection but then quit when they can't get it exactly right. For songs, too, I never have a clear vision of the final product from the start. Instead of aiming for completion, I keep thinking and creating as I go, and whatever I end up with becomes the finished piece.

メイリン　アーティスト　ソロプロジェクト「ZOMBIE-CHANG」名義で活動する。作詞作曲、ライブパフォーマンス、ボーカルなどすべてを一人で行い、自身でグッズデザイン
も行う。2022年にはアルバム『STRESS de STRESS』をリリースし、フランスツアーを決行。音楽プロジェクト以外に、モデル、執筆業などでも活動中。

Meirin is a musician who works under the name ZOMBIE-CHANG as a solo project. Meirin writes lyrics, composes music, performs live, sings vocals all by herself, and designs her own merchandise. Her latest work is the album STRESS de STRESS, which she released in 2022, followed by a tour in France. In addition to her music projects, Meirin is also active in modelling and writing.

フランスのアニエス・トゥルブレが日々の暮らしや旅先で撮影した写真を服へとプリントしたフォトコレクション。セーヌ川とパリの空が映し出されたワンピースに一目惚れ。「アニエスベーって白と黒のイメージを持たれる方が多いと思いますが、華やかなワンピースにもぜひチャレンジしてほしい」という想いから、店頭でもよく着用しているそう。首元に巻いたシルクのスカーフもアニエスベー。シンプルな組み合わせでありながら、フォトプリントを主役にした華やかさがより際立つスタイリングに。

A photo collection featuring designer Agnès Troublé's photographs, taken in her daily life and during her travels, and printed onto the clothes. Aoi instantly fell in love with this dress showcasing the Seine River and Parisian sky. 'agnès b. is often associated with black and white, but I want our customers to challenge themselves with more vibrant dresses.' The black scarf wrapped around her neck is also from agnès b. The combination is simple, but the playful print adds a touch of creativity to the styling.

ブランドスタートからアニエス・トゥルブレが大切にしてきたワークウエア。衣服として不可欠な機能性を保ちながら、色や素材を変えて新しいスタイルを生み出してきた。榎さんが着用しているのは、ピンク味のある赤色のジャケット。エネルギッシュな印象を与えつつも、柔らかい生地感なので身体にほどよく馴染んでくれる。ずっと着続けられてきたかのような着心地のよさもアニエスベーならでは。

Workwear has been important to Agnès Troublé since she started the brand. While maintaining the essential functionality of clothing, the brand has created new styles by changing colours and materials. The jacket worn by Aoi has a pinkish-red hue. It gives off an energetic impression, while the soft fabric allows it to comfortably conform to the body. The fluidity of the fabric and comfort are also distinctive features of agnès b.

Myself, Whatever My Style

どんなスタイルも、私らしくいられる

榎亜緒衣

カスタマーに寄り添い、その人にぴったりのスタイルや新しい発見をくれるのは、店舗ではたらくスタッフ。「どんなスタイルも、受け入れてくれるのがアニエスベーらしさ」と話す、アニエスベー京都BAL店 店長の榎亜緒衣。2023年にオープンしたアニエスベー祇園店の周辺を散歩しながら、4つのスタイルを紹介する。

Store staff can closely support customers and provide them with new discoveries and styles that best suit them. 'To embrace any style is what defines the essence of agnès b.' says Aoi Enoki, the manager of the Kyoto BAL store. Four styles will be introduced while Aoi strolls around the agnès b. Gion store, which opened in 2023.

花見小路を入ってすぐ、2023年4月にオープンした
アニエスベー祇園店の周辺を歩く。ワークウエアに、
艶感のあるドレッシーな素材を用いることで上品な印
象に。素材で遊びながら、本来のワークウエアの枠組
みを超えた提案も、アニエスベーの自由な発想。ジャ
ケットに異素材のブラックパンツを合わせたセットア
ップ風のスタイルが、京都祇園の街並みにも溶け込む。

Walking around the area near the agnès b. Gion store,
which opened in April 2023 just off Hanamikoji Street.
The glossy and dressy materials used on the workwear
create a delicate impression. The free-thinking of
agnès b. extends beyond the framework of traditional
workwear and offers innovative proposals while being
playful with materials. The set-up style of a jacket paired
with black pants blends seamlessly into the streetscape
of Kyoto's Gion district.

U700とよばれるアニエスベーの定番素材のワンピー
ス。1枚でさらりと着るのではなく、シャツとネクタ
イを締めてスクールガール風に着るのが榎さんらし
い。上からジャケットを羽織ったり、少しのボーイッ
シュさをプラスしたり。ガーリーやボーイッシュ、そ
の日の気分で着こなしを楽しめるのもアニエスベーの
楽しさ。

A dress made from agnès b.'s signature fabric known as
U700. Instead of wearing it alone, Aoi wore it with a shirt
and tie for a schoolgirl-inspired look. Aoi also layers a
jacket on top to add a touch of boyishness. agnès b. offers
the fun of dressing up in a girly or boyish style, allowing one
to enjoy different looks depending on the day's mood.

榎亜緒衣　アニエスベー京都BAL店 店長　アニエスベー心斎橋店で7年動めた後、2023年よりアニエスベー京都BAL店の店長に就任。エレガントからカジュアルまで自由なスタ
イリングが榎さんの特徴。アニエスベーのアイコン"レザール"に込められた意味に共感し、入社を決意した。

After working at the agnès b. Shinsaibashi store for seven years, Aoi was appointed the manager of the agnès b. Kyoto BAL store in 2023. Her unique style is
her blend of elegance and casualness. She decided to join the company because she sympathised with the meaning behind the iconic agnès b. *lézard* logo.

2023年、京都・祇園にあらたにオープンしたアニエスベー祇園店。
古くからの町屋を改装したコンセプトショップの内装を手掛けたのは、デザイナーの柳原照弘。
「光と空気」が意識された店内には京都とフランスのエスプリが融合した新しい心地よさがあった。
日本とフランスを拠点に、グローバルに活躍するデザイナーが考える土地とデザインについて。
神戸にあるスタジオ「VAGUE」を訪れ、話を聞いた。

京都から発信する
新たな文化と心地よさ
柳原照弘（デザイナー）

From Kyoto—A New Culture
and Comfort
TERUHIRO YANAGIHARA (Designer)

Photography: David Jakob
Text: Megumi Koyama

祇園という日本でも特別な文化を持つ場所。内装を手掛ける上で、どのようなところから考えていきましたか？

Y　僕はゼロから何かを生み出すのではなく、既にある状況を綿密にリサーチしたり、実際に体験したりしながら必要なものごとを考えていくタイプ。なので、単に町屋をリノベーションするのではなく、そこにアニエスベーらしいフランスのカルチャーがどう混ざり合って新しい文化をつくっていけるかということが重要でした。京都や祇園という街の歴史や紡がれてきたコンテクスト、人がどのようにその土地に根ざして生活してきたのかという部分にインスピレーションを求めたんです。

既にある歴史を引き継ぎながら、新しい文化を積み上げていく。

Y　東京やパリのアニエスベーをそのまま京都に持ってくるのではなく、京都という土地と融合していくためのプラットフォームであるべきだと思っています。なので、ブランド

としてどんなカルチャーを発信していくのかという中身の議論に時間をかけました。そうして、新しい文化を入れるにあたって必要ではないものを外していったんです。

空間に残した「必要なもの」の中で特に大切にしたものはなんでしょう。

Y　一番は中庭です。京都の中庭って特別な存在で、中に入ると庭と光が現れるすごくプライベートなつくりになっているんです。中庭だけに光が差して、そのやわらかな光が回る空間設計を行うことで、庭やそこに差す光を特別なものにして、小さな空間に奥行きをもたせているんです。

包まれるような感覚と、やわらかな光が印象的でした。

Y　一方で、フランスは屋外のダイレクトな自然光を大切にする文化がある。だから今回は、吹き抜けを中央に配

置しました。そうすることで、通常の中庭よりも光が入るようになって、店内に入ったときの空気と光が動く量を増やしています。京都とフランス、双方の光の捉え方を大切に組み立てています。

光と空気が心地よさにつながっているんですね。

Y　意識しているのは、空気をデザインするということ。そう考えると光も空気の一部ですよね。素材や設計にこだわってインテリアをデザインすることも重要ですが、それよりも大切なのが、人がそこに立ったときにどう感じるかという見えない部分。その印象や質感、やわらかさなどをすごく考えています。

左官職人が手掛けた壁や床も印象的でした。

Y　京聚楽という京都にゆかりのある素材を職人さんにつくっていただきました。壁や2階の床の一部は、元々の畳の割を活かして左官で仕上げています。左官は、土と藁、すさからできていて、水で混ぜて固めるだけで素材になり、火を入れれば焼き物にもなる。すごくシンプルな技法だけれど、いろいろな表情や機能があるんです。気候や湿度、あとは職人さんの気持ちによってすごく変化するもので、コントロールできない部分があるからこそ、対話するようにつくっていくのがおもしろいんです。

今回、内装を手掛けるにあたってパリのアニエスベーのオフィスを訪れたと聞きました。印象に残っていることはありますか?

Y　やっぱり、光ですね。美しい中庭があって、自然の石が敷き詰められた気持ちいい屋上がある。アニエスさんの部屋もそうなんです。自然の光を取り入れていて、光が差すときもあれば暗いときもある。光を通して、一日の時間を感じながらクリエイションされているんだなということを感じました。今、世界中に広がったブランドにはいろいろな側面があると思うけれど、僕が学生の頃から見ていたクリエイティブなマインドが脈々と続いている。改めてこの先どういった発信をしていくのか楽しみ

になりました。

柳原さんのデザイン哲学はどのように養われていったのでしょう?

Y　なぜ北欧やフランスのデザインが魅力的に映るんだろうって考えたとき、やっぱり歴史やその土地に根付く暮らしや文化があったんです。もちろん日本も同じだけれど、伝統=形を変えないものという意識が強くて、今現在の生活にフィットしていないことも多いと思うんです。残すべきものと、変わっていくものを見極めるという考え方は、僕自身の中に長くあるテーマ。すごくおもしろいのが、僕が手掛けたものを見て、日本の方はヨーロッパ的と言うけれど、海外の方からはすごく日本的なムードを感じると、真逆なことを言われることが多くて。表層的な部分ではなく、物事の捉え方や考え方に日本で育ってきた感性が宿っていると思うと、おもしろいですよね。

「VAGUE」と名付けたアトリエは、波を意味しているそうですね。この空間についても教えてください。

Y　日本は神戸、フランスはアルルという場所にアトリエを構えています。いろいろな国や文化を横断して仕事をしていることもあって、いつも新しい場所でどんなつながりが生まれるかということに興味があります。ものづくりをする上で、時間をともにするコミュニケーションってすごく大切だと思うんです。神戸やアルル、もちろん京都も、東京とはまた違う時間軸があって、いろいろなコミュニケーションが生まれやすい。会議室でのミーティングではなくて、ゆっくりご飯を食べて、パーソナルな部分でお互いを理解し合える余白がある。アニエスベー祇園店のプロジェクトも、はじめは京都で会って、その後もコーヒーショップでミーティングしたりしていましたね。人が集って、クリエイションの出会いや信頼関係を築くことのできる拠点。そうしてひとつの拠点から目に見えない関係性が緩やかに世界へ広がっていく。それが僕たちのスタジオの考え方だし、今回のアニエスベー祇園店にも通ずることだと思っています。

In 2023, the new agnès b. Gion store opened in Kyoto's Gion district. The concept shop, housed in a renovated *machiya* (traditional Japanese townhouse), was designed by Teruhiro Yanagihara. Inspired by 'light and air', the interior offers a new sense of comfort that blends the essence of Kyoto and France. We visited his studio, Vague, in Kobe to discuss the relationship between place and design with Teruhiro, a globally active designer with bases in Japan and France.

Gion in Kyoto is a place with a unique cultural heritage in Japan. How did you approach designing the shop's interior?

Y　Rather than creating something from scratch, I am the type of person who meticulously researches and immerses himself in existing conditions to understand what is needed. Therefore, it was important not just to renovate the *machiya*, but to consider how the French culture of agnès b. could blend to create a new cultural experience. I drew inspiration from the history and context of Kyoto and Gion and from understanding how people have lived and rooted themselves in this place over time.

You are creating a new culture while inheriting the existing history.

Y　Instead of simply bringing the agnès b. experience from Tokyo or Paris to Kyoto, it was important to create a platform that would merge with the unique essence of Kyoto. Therefore, we spent a significant amount of time discussing the cultural message the brand wanted to convey. In the process, we carefully removed unnecessary things to incorporate the new cultural elements.

Which of the essential elements you retained in the space was the most important?

Y　The most crucial element is the courtyard. Courtyards in Kyoto are unique; once you enter, you are greeted with a private space where the garden and light appear. By designing the space so that light enters the courtyard and diffuses softly into the space, the garden and the light falling on it became something special and helped to add depth to the small space.

The sense of being enveloped and the soft light in the new shop left a deep impression.

Y　France has a culture that values direct natural light outdoors. This is why we placed an atrium in the centre this time. This allows more light to enter than a typical courtyard, increasing the movement of air and light when you enter the store. We carefully integrated both Kyoto's and France's approaches to capturing light.

Both the light and air contribute to a sense of comfort.

Y　I focus on designing the air, and light is also part of it. While it is important to design the interior with a focus on materials and structure, what is even more crucial is the intangible aspect of how people feel when they stand in that space. We put a lot of thought into the impressions, textures, and softness that people experience.

The walls and floors crafted by plaster artisans were also impressive.

Y　We had artisans create work on the surfaces using *kyōjuraku*, a traditional material from Kyoto. Some of the walls and parts of the second floor are finished with plaster, incorporating the original tatami dimensions. Plaster, made from earth, straw, and fibre, is mixed with water to harden and can also be fired to become pottery. It's a very simple technique, but it offers a variety of textures and functions. The result can vary significantly with the climate, humidity, and even the artisan's mood, making it an interesting process that involves dialogue with the material.

You visited the Paris office while working on the interior design this time. Was there anything that left a lasting impression?

Y　It was the light. There is a beautiful courtyard and a comfortable rooftop paved with natural stone. Agnès' room is the same. It incorporates natural light that changes throughout the day—sometimes bright or sometimes dark. Through the light, there is a sense of time passing, which has probably influenced her creative process. Although many different aspects of the brand have spread worldwide, the creative spirit I saw when I was still a student continues to thrive. I am excited to see how the brand will continue to evolve and what new expressions will emerge.

How did you develop your design philosophy?

Y　When I thought about why Scandinavian and French designs are so attractive, I realised it's because of the history and the culture deeply rooted in their daily lives. Of course, Japan has the same qualities. However, there is a strong perception that tradition equals something unchanging, which often doesn't fit with contemporary life. The idea of distinguishing between what should be preserved and what should change and evolve has been my long-standing theme. Interestingly, Japanese people often describe my work as European, while people from abroad often sense a distinctly Japanese atmosphere. It's fascinating because it shows that my perception and thought processes, shaped by the sensitivities from growing up in Japan, are embedded in my design beyond the surface elements.

The name of your atelier, 'Vague', means waves. Could you tell us more about this space?

Y　We have ateliers in Kobe, Japan, and Arles, France. Since my work spans various countries and cultures, I am always intrigued by the connections that emerge in new places. In the process of creating, communication through shared time is extremely important. Kobe, Arles, and of course, Kyoto all have a different time flow than Tokyo, making it easier to facilitate various forms of communication. Instead of having meetings in a conference room, we take our time, share meals, and create space to understand each other on a personal level. The project for the agnès b. Gion store also started with a meeting in Kyoto, followed by discussions in coffee shops. It's a place where people can gather and build creative encounters and trust. These invisible connections can gradually spread into the world from a single hub. This is the philosophy of our studio, and it applies to the agnès b. Gion store project as well.

柳原照弘　デザイナー　Teruhiro Yanagihara Studio 主宰。インテリアデザイン、プロダクトデザイン、クリエイティブディレクションなど包括的な提案を行う。神戸とフランスのアルルに自身のスタジオ「VAGUE」を構え、空間を可視化するフレグランスブランド〈LICHEN〉をはじめとしたプロジェクトや、ものづくりの思考やプロセスを表現・発信する場として活動を広げている。

Teruhiro Yanagihara is a designer and founder of Teruhiro Yanagihara Studio, an interior design studio that provides comprehensive proposals, including interior design, product design, and creative direction. His studio, VAGUE has locations in Kobe and Arles in France—it is a space to visualise and communicate his design concepts, thinking, and creative process through projects such as the original fragrance brand LICHEN.

The Fusion of Cuisines from France and Kyoto

おいしさの秘密

渥美彰人

日本上陸40周年を迎えた2023年に、京都に新しくオープンしたアニエスベー祇園店。フランスと京都の文化が融合する建物の1階には、カフェが併設されている。メニューを考案したシェフの渥美彰人は「ここで自分が食べたいものはなんだろう？」そんな素直な感情を起点にアイデアを膨らませたと話す。フランスでキャリアをスタートし、京都を拠点に活躍する彼の視点から生まれた色鮮やかで繊細な味わいのタルティーヌやサンドウィッチ。料理との出会いやメニューにまつわる話を聞いた。

In 2023, agnès b. celebrated its 40th anniversary in Japan with the opening of a new store in Gion, Kyoto. The building beautifully merges French and Kyoto cultures, and the first floor features an adjoining café. Chef Akihito Atsumi, who designed the menu, explains that he expanded his ideas from the simple question, 'What would I want to eat here?' Having started his career in France and now working in Kyoto, Akihito's unique perspective brings to life vibrant and delicate flavours in his tartines and sandwiches. We had the pleasure of discussing his culinary journey, the inspirations behind the menu, and his heartwarming, satisfying dishes.

料理の世界に魅了された漫画との出会い
料理に興味を持ち始めた9歳の頃、それを知った母の友人が料理漫画を教えてくれたんです。寺沢大介さんの『将太の寿司』と『ミスター味っ子』。このふたつの漫画で描かれる料理の世界に魅了されて、どんどんのめり込んでいきました。当時は卵焼きやチャーハンなどをよく作っていましたね。小さい頃から料理に興味を持っていたので、シェフを目指すのも自然な流れでした。

An encounter with mangas sparked fascination for the world of cooking
When I was nine and just starting to develop an interest in cooking, a friend of my mother's introduced me to cooking manga. It was Daisuke Terasawa's *Shōta no Sushi* and *Mister Ajikko*. I was captivated by the culinary world depicted in these two mangas and became increasingly absorbed in it. Back then, I often made *tamagoyaki* (Japanese rolled omelette) and fried rice. Having been interested in cooking from a young age, it was a natural progression for me to aspire becoming a chef.

アルザスで学んだフランス料理
留学中はフランスのアルザス地方にある、現地のレストラン「オーベルジュ・ド・リル」で修行を積みました。アルザスは北東に位置しておりドイツの国境に程近く映画『ハウルの動く城』の舞台にもなっています。その中でも、ストラスブール近郊のイローゼルンという村に滞在していました。1日にバスが2本しか走らないような田舎で、フランスとドイツ、両方の文化が交差する歴史的背景を持つ場所です。自然豊かな場所で地元の食材を用いた伝統と現代のフランス料理を学びました。

Learning French cuisine in Alsace
During my studies abroad, I trained at a local restaurant called 'L'Auberge de l'Ill' in the Alsace region of France. Alsace is located in the northeast, near the German border, and is also the setting for the movie *Howl's Moving Castle*. I stayed in a town called Illhaeusern, near Strasbourg. It's a rural area with only two buses running each day, and it has a rich history where French and German cultures intersect. In this nature-rich setting, I learned about both traditional and contemporary French cuisine, using local ingredients.

自分の個性を活かすために

フランス料理の道へと進んだのは、「左利き」だったこと
が大きな理由です。日本料理はどうしても右利きの盛り
付け方や作法、器具が多いんです。まだまだ右利きに
矯正される時代だったのですが、そこに時間を割くより
も自分の出来ることを伸ばしたいと思ったんです。調理
師学校では、フランス料理を専攻し、本場の料理を学
ぶためフランス校に留学をしました。

Embracing my uniqueness

A significant reason I pursued French cuisine was
because I am left-handed. Japanese cuisine often
involves right-handed plating techniques, methods,
and tools. At a time when left-handedness was still
usually corrected, I wanted to focus on enhancing
what I could do rather than spending time adapting to
right-handed practices. I majored in French cuisine in
culinary school and studied abroad in France to learn
authentic French cooking.

フランスの国民食と京都ならではの味わい

メニュー開発ではフランスの食文化、そして日本の中でも京
都の食文化を良いバランスで融合させていくことを大事に考
えました。フランスの暮らしに根付いたものをと思い、タル
ティーヌやクロックムッシュ、カスクルートなどの国民食を取
り入れています。味に多様性があることもフランス料理の特
徴です。バターはもちろん、スパイスなどではの味の幅がとても広く、
今回はそれらと相性の良い京都ならではの関西の白味噌を
ベースに、フランス風のドレッシングを掛け合わせたものを
料理に用いています。今後も積極的に私なりの解釈を加え、
京都とフランスの食文化を融合させた料理にチャレンジして
いきたいと思っています。

French national dishes and Kyoto's unique flavours

In developing the menu, I focused on striking a good balance
between French culinary culture and the unique food culture of
Kyoto. I incorporated French national dishes such as tartines,
croque-monsieur, and casse-croûte, which are deeply rooted
in French daily life. One characteristic of French cuisine is its
diversity of flavours. French cuisine often features multiple
layers of taste, including butter and a variety of spices. This
time, I used Kyoto's unique Kansai-style white miso as a base,
pairing it with French-style dressings to create a harmonious
blend of flavours. In the future, I aim to continue integrating my
interpretations, blending the culinary cultures of Kyoto and
France in new and exciting ways.

メニューは頭の中に
ひらめきは暮らしの中に

メニューは頭の中で考えることが多いのですが、最初のイン
スピレーションは日々の暮らしの中にあります。山の表情か
ら季節の移り変わりを感じて、そろそろこの食材が食べ頃だ
なとか。ちょっと寒くなってきたから、スープを作ろうかなと
いった風に繋がっていく。そういう意味では、京都は自然が
すぐそばにあり生産者さんとの距離も近いので、その時の旬
をキャッチできる環境が料理に最大限活かされています。

Menus in my mind,
inspiration from daily life

I often develop menus in my mind, but the initial inspiration
comes from daily life. Observing the mountains and sensing
the change of seasons, I can tell when certain ingredients are
at their peak. When it starts to get colder, I might think about
making soup. In this sense, Kyoto, with its proximity to nature
and close relationship with local producers, provides an ideal
environment for capturing the essence of seasonal ingredients
and incorporating them into my cooking.

屋美彰人　シェフ　辻調理師専門学校・辻調フランス校卒業。在学中に仏アルザス地方のレストラン「オーベルジュ・ド・リル」にて修行を積む。卒業後は東京の「L'Effervescence」
「TIRPSE」を経て、現在は京都の「cenci」にてスーシェフをつとめる。

Akihito Atsumi graduated from Tsuji Culinary Institute and Tsuji French School. While still a student, Akihito trained at the restaurant 'L'Auberge de l'Ill' in Alsace, France. After graduation

2023年、アニエスベー祇園店のオープンを記念して公開されたスペシャルムービーと、
アニエスベー京都BAL店にて披露された一度限りのライブパフォーマンス。
踊るだけでなく、それらの演出まで手がけたダンサー、俳優のアオイヤマダ。
アニエスベーの為に特別に制作されたムービーと
ダンスパフォーマンスのバックステージをドキュメントしながら、
彼女のインスピレーションとダンスを通して伝えたいメッセージを聞いた。

心とからだが一つになる時

アオイヤマダ（ダンサー, 俳優）

When Mind and Body Become One
AOI YAMADA (Dancer, Actor)

Photography: Shota Kono
Text: Megumi Koyama

「規則や決まりの中の境界を常に問いかけてきました。制服や靴、髪型も決められた校則の厳しい中学校に通っていたんですが、なぜか雨の日の長靴は自由だったんです。じゃあこの"長靴"って何を指すんだろうって。自分なりに考えて、厚底の真っ黒な靴を履いたり。男子が禁止されていたツーブロックにして、女子だったらどうなの？って。みんな違和感を持たずにルールに従うけれど、そうやって私が突っつくことによって、考え直すきっかけになりますよね。それってハプニングでもあるし、とても楽しかった。規則に縛られていた中学生時代は、自分の表現というものを広げようと必死だったのかもしれない。けれど同時に、私にとって表現するきっかけをくれたような気がしています」

ダンスパフォーマンスは、鑑賞者だけでなくパフォーマーにとっても何が起こるかわからないハプニングの連続。かつて彼女の校則への問いかけがハプニングであったように、今ではダンスを通して社会や鑑賞者に問い続ける。そんな彼女の創作の源のひとつに、幼少期の母親との思い出があるという。

「母はとても自由な人で、身の回りにあるものに手を加えて創造する楽しさを教えてくれました。例えば、折り紙でおままごとセットを作るとか、落ちている石に絵を描いて宝石に見立てたりとか。遊び道具を買ってもらうのではなく、いつも自分たちで遊びを作っていました。そんな発想から影響を受けて生まれたのが"野菜ダンス"。コロナ禍に、祖母に元気だよと伝えるために作ったのがはじまりでした。言葉、野菜、そして祖母が知っている歌謡曲。身の回りにあるこれらを組み合わせたらなんだか新しくて、楽しかった」

言葉や音楽、野菜など、身近な物事をはじめ、旅先での景色や人との会話から生まれる感情も大切なインスピレーションのひとつだという。

「抽象化された感情に自分自身の身体を近づけていく。感情と身体がひとつになる瞬間が私はダンスだと思っています。だから、感情の種類が増えるほど、動きの幅が増えていく。エジプトに行った時に砂漠の上を板に乗って滑ったときのこと。細

agnès b. stories Interview

agnès b. stories. Interview.

かい砂が肌を突き抜けていく感覚や、光のない中でキャンプをして、そこで遊牧民が焼いてくれた鶏肉を食べた時の安心感。車に揺られながら、砂漠の中をひたすら走り続ける時の不安など、今までにない感情と出会った。その感情を思い出した時、新たな動きが生まれるのを身体で感じました」

その場を温かいエネルギーで包み込み、一瞬にして雰囲気を変えてしまうアオイヤマダさんのパフォーマンス。その裏側には、パフォーマンスをする場所やそこで感じた空気、衣装、音楽など様々な要素に一つひとつ向き合いながら、時には観客の視点に立ち、丁寧に作品を作り上げていく姿があった。その一方で、難しいことは考えずに自由に表現を楽しむことも大切にしているという。

「新しいものや、自分しかできないものを追い求めて探そうとすると苦しくなってしまうことがあります。特に私たちの世代は、それにつまずいてしまう人が多い。でも、冷蔵庫の中にある野菜と最近ハマってるアイドルの曲を組み合わせたら『あれちょっと面白いかも?』とかそういうことでいいと思うんです。もっと好きなものを自由に楽しんでほしい。発想の種は足元に落ちていて、ほらなんか新芽が出ているかもよ、っていうポジティブなメッセージを伝えたいです」

In 2023, to commemorate the opening of the agnès b. Gion store, a special movie was released, and a one-time live performance was showcased at the agnès b. Kyoto BAL store. Aoi Yamada directed these performances and contributed as a dancer. This time, while documenting the backstage of the specially created movie and dance performance for agnès b., we asked about her inspiration and the message Aoi wants to convey through her dance.

'I have always questioned the boundaries within rules and regulations. I attended a strict junior high school with specific rules for uniforms, shoes, and hairstyles. However, for some reason, rain boots were the only exception. So, I wondered, what exactly do "rain boots" refer to? And I wore thick-soled, all-black shoes. I also tried a two-block haircut, which was prohibited for boys, thinking, "What about for girls?" Everyone follows the rules without question, but by challenging them, I could make people reconsider why. My time in high school, constrained by rules, might have been a period where I desperately tried to expand my own form of expression. At the same time, it gave me the opportunity to explore and express myself.'

At the agnès b. Kyoto BAL store, Aoi performed a piece themed around the beloved stripes that have been a favourite since the brand's establishment. The head accessory featuring the brand's iconic *lézard* was specially created for this occasion.

Dance performances are a series of unpredictable events, not just for the audience but for the performer as well. Just as her self-expression and questioning of school rules once created disruptions, Aoi now continues to challenge society and audiences through her dance. One of the sources of her creativity is her childhood memories with her mother.

'My mother was a very free-spirited person and taught me the joy of creating things with what we had around us. For example, we made playsets out of origami or painted stones to make them look like jewels. Instead of buying toys, we always created our own games. One of the ideas influenced by this creativity was the "Vegetable Dance". It started during the pandemic as a way to show my grandmother that I was doing well. By combining words, vegetables, and songs my grandmother knew, it turned into something new and fun.'

Aoi considers words, music, and vegetables, as well as emotions arising from the sights on her travels and conversations with people, to be important sources of inspiration.

'I bring my body closer to abstracted emotions. I believe dance is the moment when these emotions and the body become one. Therefore, the more types of emotions I experience, the greater the range of movement I can do. When I went to Egypt, I slid across the desert on a board. The sensation of fine sand piercing my skin, camping in the darkness, and the comforting feeling of eating chicken cooked by the nomads…The unease of driving endlessly through the desert—all these were new emotions for me. When I recall these emotions, I can feel new movements emerging in my body.'

While engaging in the abstract expression of dance, the process of concretely visualising ideas is also essential. In addition to dancing, Aoi has increasingly taken on the role of directing, as seen in this performance. Aoi diligently works through a notebook filled with notes right up until the performance.

Aoi's performances envelop the space with warm energy, instantly transforming the atmosphere. Behind the scenes, she meticulously considers each element—the performance venue, the ambiance, the costumes, and the music—while sometimes adopting the audience's perspective to create her pieces with care. At the same time, Aoi values the freedom to enjoy expressing herself without overthinking.

'Trying to chase after new things or something only I can do can sometimes become overwhelming. Many people in our generation struggle with this. But I think it's perfectly fine to combine the vegetables in your fridge with a song from your current favourite idol and think, "Hey, this might be interesting." I want people to freely enjoy what they love. The seeds of ideas are right at your feet, and look, they might just be sprouting. That's the positive message I want to convey.'

ダンスという抽象的な表現を行う一方で、具体的に可視化することも欠かせない制作プロセス。今回のパフォーマンスのように踊るだけでなく、演出を手がけることも増えてきたという。パフォーマンス直前までメモ書きがぎっしり詰まったノートと向き合う姿が印象的だった。

While engaging in the abstract expression of dance, the process of concretely visualising ideas is also essential. In addition to dancing, Aoi has increasingly taken on the role of directing, as seen in this performance. Aoi diligently works through a notebook filled with notes right up until the performance.

アニエスベー京都BAL店ではブランド設立当初から愛され続けるボーダーをテーマにパフォーマンスを披露した。ブランドのアイコンであるレザールのヘッドアクセサリーは、この日のために特別に制作されたもの。

At the agnès b. Kyoto BAL store, Aoi performed a piece themed around the beloved stripes that have been a favourite since the brand's establishment. The head accessory featuring the brand's iconic *lézard* was specially created for this occasion.

アオイヤマダ　ダンサー，俳優　東京2020オリンピックの閉会式でのソロパフォーマンスをはじめ、アーティストのMV、Netflixドラマや舞台に出演。ダンスという表現を活かしながら、枠を超えてモデルや表現者として媒体を問わず活躍。彼女を筆頭としたクリエイティブチームEBIZAZEN[海老坐禅]など、自身で音楽やダンスのクリエイションに携わる。

From her solo performance at the Tokyo 2020 Olympics closing ceremony to appearances in artists' music videos, Netflix dramas, and stage productions, Aoi Yamada transcends boundaries, utilising dance as a form of expression while thriving as a model and performer across various media. Aoi is also involved in music and dance creation with her creative team EBIZAZEN, leading the way in innovative artistic projects.

AGNÈS TROUBLÉ

hello dear !! japan !! ♡ 40 years after agnès

アニエス・トゥルブレからの手紙

Letter from Agnès Troublé

アニエス・トゥルブレ　デザイナー　フランス・ヴェルサイユ生まれ。エディターを務めたのち、1975年にアニエスベーを立ち上げ、1983年に日本上陸。1984年、ギャラリー デュ ジュールをオープン。ブランド設立から現在まで、デザイナーとして活動を続けている。

Agnès Troublé was born in Versailles, France. After working as an editor, Agnès founded agnès b. in 1975 and expanded to Japan in 1983. In 1984, the art gallery Galerie du jour was opened in Paris. Since the establishment of the brand, Agnès has continued to work as a designer.

出演者、フォトグラファー、この本に関わってくれたすべての人たちに感謝します。

アニエスベー

Special thanks to the cast, photographers and everyone involved in this book.

agnès b.

アートディレクション・デザイン：前田晃伸（MAEDA DESIGN LLC.）
デザイン：庄野祐輔、黒木晃（MAEDA DESIGN LLC.）
編集・テキスト：小山めぐみ、ヴィクター・ルクレア（株式会社 コンタクト）
プロジェクトマネージメント・テキスト：神田春樹（株式会社 コンタクト）
プロジェクトマネージメント：遠藤慶太（株式会社 コンタクト）
テキスト：小川知子、堺れい、原智慧
翻訳：ジョイス・ラム
進行：新庄清二（株式会社青幻舎）
協力：細田真菜葉、難波侑希（株式会社 ウィークデー）

agnès b. stories　アニエスベー ストーリーズ

発行日：2024年7月29日　初版発行
監修：アニエスベー

発行者：片山誠
発行所：株式会社青幻舎
京都市中京区梅忠町9-1　〒604-8136
Tel.075-252-6766 Fax.075-252-6770
https://www.seigensha.com

印刷・製本：株式会社山田写真製版所

Art Direction, Design: Akinobu Maeda (MAEDA DESIGN LLC.)
Design: Yusuke Shono, Akira Kuroki (MAEDA DESIGN LLC.)
Editing, Text: Megumi Koyama, Victor Leclercq (kontakt co., ltd.)
Project Management, Text: Haruki Kanda (kontakt co., ltd.)
Project Management: Keita Endo (kontakt co., ltd.)
Text: Tomoko Ogawa, Rei Sakai, Chikei Hara
Translation: Joyce Lam
Production Management: Seiji Shinjo (Seigensha Art Publishing, Inc.)
Cooperation: Manaha Hosoda, Yuki Namba (Weekday, Inc.)

agnès b. stories

First Edition: 29 July, 2024
Supervision: agnès b.

Publisher: Makoto Katayama
Published by Seigensha Art Publishing, Inc.
9-1, Umetada-cho, Nakagyo-ku, Kyoto, 604-8136, Japan
Tel: +81 75 252 6766 Fax: +81 75 252 6770
https://www.seigensha.com

Printed and Bound by: Yamada Photo Process Co., Ltd.

1975

Opening of the first agnès b. shop
アニエスベー1号店オープン

アニエスベーは最初のショップをパリのレアール地区にある古い肉屋を全面改装して開業しました。当時、その地区は急激に発展している最中でした。オフィスやアトリエとしてだけでなく、表現の場、あるいは出会いの場として使われました。店内には、グラフィティや様々なポスターが貼られ、鳥が自由に飛び回り、お客さんやスタッフたちが座って会話を楽しみ、音楽も聴くことができました。

Fully renovated from an old butcher shop, agnès b. first opened its shop in Les Halles district in Paris. At that time, the district was in the midst of rapid development. The space was not only used as an office or atelier, but it was also a place for expression and socialising. Inside the store, graffiti and different kinds of posters were displayed, and birds flew freely around. Customers and staff could sit, enjoy conversations, and listen to music.

1979

Launch of the snap cardigan
カーディガンプレッション誕生

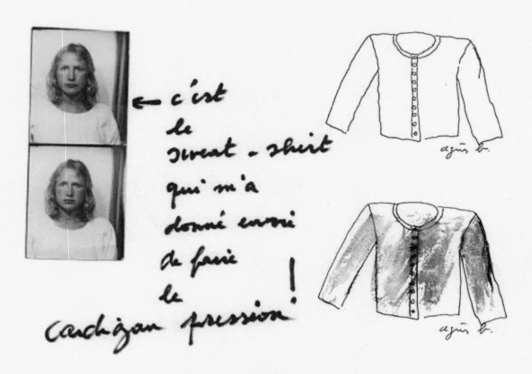

← c'est
le
sweat - shirt
qui m'a
donné envie
de faire
le
cardigan pression !

「随分前に、自分用にカーディガンをデザインしました。スナップボタンが沢山並んだ18世紀の洋服のような前開きのスウェットで、大人のための子供服、またはその逆のような服を作りたかったのです」。ブランド創始者のアニエス・トゥルブレが生み出したカーディガンプレッションは、1979年の登場以来永く愛されています。

'I designed a cardigan for myself quite a while ago. It was an open-front sweatshirt, like a garment from the 18th century, with lots of snap buttons on it. I wanted to create clothes that were like children's clothing for adults, or perhaps the other way around.' Since its debut in 1979, the snap cardigan—created by the founder of the brand, Agnès Troublé—has been loved for many years.

1983

Launch of agnès b. in Japan
アニエスベー日本上陸

東京にオープンしたアニエスベーのブティックは「東京
の家」と呼ばれるほどアニエス・トゥルブレ本人が愛す
場所となりました。当時、ボーダー T シャツやカーデ
ィガンプレッションをはじめとした、シンプルでフレン
チシックなアイテムが日本で大流行しました。

The agnès b. boutique that opened in Tokyo became a
beloved place for Agnès Troublé herself, to the extent
that she referred to it as her 'home in Tokyo'. At that
time, simple and French-chic items like the striped
T-shirt and snap cardigan became extremely popular
in Japan.

1984

Opening of Galerie du jour
ギャラリー デュ ジュール オープン

「私は自分の好きなものを見せるギャラリーを作り
たかったのです。ギャラリーと呼んでいるけれど、
ペインティング、彫刻、写真、そして、物事の色々
な側面を見せるための場所、といえるのではないで
しょうか。同時に毎回、下絵、ステンシル、スクリ
ーンプリント、版画など、イメージを拡散する新し
い方法を考え出して、皆がそれを手にできるように
したいと思っています。またギャラリー主催の展覧
会も行います。場所はまだきれいではないし、肉屋
の面影が残っていますが、人々がまた来たいと思っ
て集まる活気のある場所になればと思っています」。
1984年のオープン以来、ギャラリー デュ ジュー
ルは現代アートギャラリーとして多くのアーティス
トの作品を紹介しています。

'I wanted to create a gallery to showcase the
things I love. Although it was called a gallery,
it was a place to show paintings, sculptures,
photographs and the different aspects of different
things. At the same time, I always think of new
ways to disseminate images, such as sketches,
stencils, screen prints, and engravings, so that
everyone can hold onto something from the
exhibitions. We would also host exhibitions
organised by the gallery. The place was not clean
yet, and it had traces of the butcher's shop, but I
hoped that it would become a lively place where
people would want to come back.' Since its opening
in 1984, Galerie du jour has introduced the works
of many contemporary artists.

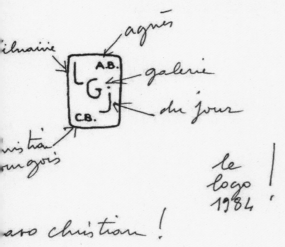

1987 Opening of the first agnès b. shop in London
ロンドンに初のショップがオープン

1992 Featured in Quentin Tarantino films
クエンティン・タランティーノ作品への衣装提供

映画監督のクエンティン・タランティーノの作品『レザ
ボア・ドッグス』に衣装提供をしました。映画を象徴す
る黒いスーツはアニエスベーのものでした。

Provided costumes for actors in the film *Reservoir Dogs* by Quentin Tarantino. The iconic black suit in the film was designed by agnès b.

1994

Launch of Artist T-shirts
アーティストTシャツ発表開始

キューバ人アーティストのフェリックス・ゴンザレス＝トレスと出会い、初めてのアーティストTシャツを制作しました。

The first Artist T-shirt was created in collaboration with Cuban artist Félix González-Torres.

1997 Start of *Le Point d'Ironie*
『ポワンディロニー』創刊

『ポワンディロニー』はアニエス・トゥルブレと現代美術家のクリスチャン・ボルタンスキー、現代美術キュレーターのハンス＝ウルリッヒ・オブリストとのある会話をきっかけに生まれました。サイズや配布方法などもユニークな無料の定期刊行物で、"分散"というアイデアのもと、毎号10万部を世界中のアニエスベーブティック、美術館、ギャラリー、学校、カフェ、映画館など様々な場所で配布しています。

Born from a conversation between Agnès Troublé, contemporary artist Christian Boltanski, and contemporary art curator Hans Ulrich Obrist. This free periodical, unique in its size and distribution methods, is created under the idea of 'dispersal'. With a print run of 100,000 copies for each issue, the magazine is distributed free of charge at various locations worldwide, including agnès b. boutiques, museums, galleries, schools, cafés, and cinemas.

2006

30th anniversary of agnès b.
メゾン30周年

メゾン30周年を記念して、パリのオランピアで音楽ライブを開催。2007年春夏レディースコレクションショーに続いて、由緒ある音楽ホールにロックンロールの音が鳴り響きました。ライブには、パティ・スミス、ジャン=ルイ・オベール、プラシーボ、ザ・ヴァージンズ、ソニック・ユース、スプリット、ポニ・ホックス、ラファエルなど、伝説的なロックミュージシャンから、新進気鋭のアーティストまでが集結しました。

To celebrate the brand's 30th anniversary, a music live event was held at Olympia in Paris. Following the Spring/Summer 2007 women's collection show, the historic Paris music hall resonated with the sounds of rock'n'roll. The live event brought together legendary rock artists and emerging talents, including Patti Smith, Jean-Louis Aubert, Placebo, The Virgins, Sonic Youth, Split, Poni Hoax, Raphael, and others.

2009 Launch of 'To b. by agnès b.'
〈トゥー ビー バイ アニエスベー〉ローンチ

〈トゥー ビー バイ アニエスベー〉は、アニエス・トゥルブレがこよなく愛し、デザインするのが楽しくてたまらないという、女の子のための洋服。日本の女の子たちをインスピレーションに誕生した、セカンドラインです。

A second line inspired by Japanese girls was launched. Agnès Troublé immensely loves and enjoys designing clothes for young girls.

2016 Establishment of the Tara Océan Foundation
タラ オセアン財団設立

アニエスベーフランス本社のCOOであり、Tara Océan財団共同創設者兼会長であるエティエンヌ・ブルゴワが、アニエス・トゥルブレと共に支援しているプロジェクトのための財団を設立。科学的、環境保護的ミッションを遂行するため、2003年にスクーナー船を購入し、タラ号と名付けました。以来、科学探査船タラ号は世界中を航海し、海が直面する環境的脅威や気候変化の影響を調査しています。

A project supported by Étienne Bourgois, COO of agnès b. France headquarters and co-founder & chairman of the Tara Océan Foundation, together with Agnès Troublé. They purchased a schooner, which was named Tara, to carry out scientific and environmental missions. Tara, the scientific research vessel sails around the world, investigating the environmental threats and impacts of climate change on the oceans.

2019

40th anniversary of the snap cardigan
カーディガンプレッション誕生40周年

1979年の夏の暑い日、アニエス・トゥルブレは着ていた白いスウェットシャツの前身頃をハサミで切り開き、スナップボタンをルネッサンス調の服のようにずらりと並べてカーディガンに作りかえました。それ以来、このカーディガンプレッションは、素材や形を変えて進化しながら、メンズ、レディース、キッズ、ベビーまで世界中の様々な世代から愛されるタイムレスなロングセラーアイテムに。カーディガンプレッションの写真集が発売されたほか、カーディガンプレッションをテーマに、パリでは映画監督のデヴィッド・リンチや現代アーティストのアネット・メサジェなど、世界的に著名なクリエーターたちによる写真展が。アニエスベー青山店ではHIROMIXが様々なモデルのカーディガンプレッションスタイルを撮影した写真展、青山のアニエスベー ギャラリー ブティックでは川内倫子が撮りおろした写真展「When I was seven.」が開催されました。

On a hot summer day in 1979, Agnès Troublé cut open the front of the white sweatshirt she was wearing with scissors and arranged snap buttons in a Renaissance style to transform it into a cardigan. Since then, this snap cardigan has evolved in materials and shapes, becoming a timeless, long-selling item loved by generations worldwide, from men and women, to children and babies. In addition to the release of a photo-book on the snap cardigan, a photo exhibition themed around the snap cardigan was held in Paris, featuring globally renowned creators such as film director David Lynch and contemporary artist Annette Messager. At the agnès b. Aoyama store, a photo exhibition by HIROMIX showcased photos of various models of the snap cardigan, and in the agnès b. galerie boutique Tokyo, Rinko Kawauchi's photo exhibition *When I was seven.* was held.

2019

Opening of the agnès b. Shibuya shop
アニエスベー渋谷店オープン

2023

40th anniversary of agnès b. in Japan
日本上陸40周年

アニエスベーにとって特別な場所である京都。かつてアニエス・トゥルブレが訪れ、以来こよなく愛することとなった京都の祇園に、町屋を一軒丸ごと改築したブランドの新たなコンセプトショップが誕生しました。デザイナーの柳原照弘が手掛けたショップは、1階はカフェ、2階はコレクションや限定商品の展示・販売。1階のカフェでは、「Overview Coffee」による淹れたてコーヒーを味わえるほか、アルザスのレストラン「L'Auberge de l'Ill」にて修行を積み、現在は京都の「cenci」でスーシェフを務める渥美彰人によるフードメニューや、京都のベーカリーカフェ「Camphora」を主宰するパティシエ 丸橋理人によるデザートメニューなどを提供しています。

Kyoto is a special city for agnès b. In Gion, Kyoto—a place Agnès Troublé visited and has deeply loved ever since—a brand new concept shop has been established in a fully renovated *machiya* (traditional Japanese townhouse). Designed by Teruhiro Yanagihara, the shop features a café on the first floor and showcases collections and limited edition items on the second floor. The café on the first floor offers freshly brewed coffee by 'Overview Coffee', as well as food developed by Akihito Atsumi, who trained at 'L'Auberge de l'Ill' in Alsace and is currently the sous-chef at 'cenci' in Kyoto. Desserts are provided by pastry chef Rihito Maruhashi, who runs the bakery café 'Camphora' in Kyoto.

Chronicle of agnès b.

1941年	・11月26日 ヴェルサイユにてアニエス・トゥルブレ誕生
1958年	・クリスチャン・ブルゴワ氏と結婚 ・パリのケルベール通りにあるジャン・フルニエ・アートギャラリーにて働きはじめる
1962年	・アニエスベーが『エル』誌に初めて掲載された際、アニエスがブランドロゴを考案する
1966年	・映画『ポリー・マグーお前は誰だ』に登場した有名な黒と白のボーダーTシャツをデザイン
1973年	・アニエスベー ブランド商標登録
1975年	・パリのレアール地区 ジュール通り3番地に店舗オープン
1978年	・パリのミシュレ通りに、1910年に作られた食料品店を改装した板張りの新店舗をオープン
1979年	・カーディガンプレッション誕生
1980年	・ニューヨークのソーホーに店舗をオープン
1981年	・メンズ1号店がジュール通り3番地にオープン ・レディースとキッズの店舗がジュール通り2番地に移転
1983年	・アニエスベー、日本上陸
1984年	・ギャラリー デュ ジュール オープン
1985年	・国家功労賞勲章シュヴァリエ賞をフランソワ・ミッテランから授与される
1986年	・ジャン＝バプティスト・モンディーノ、ジル・ベンシモン、ドミニク・イッセルマン、ジャンルー・シーフら64人の写真家を起用し、カーディガンプレッションをテーマに撮影。作品は同年にギャラリー デュ ジュールで展示
1987年	・ロンドンのフルハム通りに店舗オープン ・ロレアルのクラブ・クレアター・ボーテと化粧品ラインをローンチ ・同年、香水〈ル・ベー〉を発表
1988年	・ブルース・ウェーバー監督作品『レッツ・ゲット・ロスト』のチェット・ベイカーに衣装提供
1992年	・クエンティン・タランティーノ監督作品『レザボア・ドッグス』に衣装提供 ※アニエスベーのファンであるアメリカ人監督のタランティーノは『パルプ・フィクション』(1994)でユマ・サーマンとジョン・トラボルタの衣装にもアニエスベーの洋服を採用
1993年	・バッグと革小物のライン〈アニエスベー ボヤージュ〉誕生
1994年	・キューバ人アーティスト、フェリックス・ゴンザレス＝トレスと出会い、初めてのアーティストTシャツを制作

| 1995年 | ・中国、香港に上陸 |
| | ・エイズ撲滅運動のために、店舗で無料コンドームの配布を決定 |

| 1996年 | ・パリ、ポンピドゥー・センターにて「写真家とカーディガンプレッション」展開催 |

1997年	・国家功労勲章オフィシエ賞を受賞
	・デヴィッド・ボウイのニューヨーク、マディソン・スクエア・ガーデンでのコンサートに衣装提供
	・アニエスと現代美術家 クリスチャン・ボルタンスキー、現代美術キュレーター ハンス=ウルリッヒ・オブリストとの会話をきっかけに『ポワンディロニー』が誕生

| 1998年 | ・ギャラリー デュ ジュールがジュール通りから、パリ4区のカンカンポワ通り44番地に移転 |

| 2000年 | ・レジオン・ドヌール勲章シュヴァリエ賞を受賞 |
| | ・アメリカのブランド、エバーラストとのコラボレーションによる、〈ビー・エバーラスト〉ライン誕生 |

| 2001年 | ・香港にアニエスベー ギャラリー ブティックがオープン |

| 2003年 | ・科学探査船を購入しタラ号と名付け、世界中の海をよりよく理解するために探査し、その結果を共有するタラ号プロジェクトを始める |

| 2006年 | ・メゾン30周年を記念し、オリンピアでファッションショーとコンサート開催 |
| | ・創業2年目のフランスのスニーカーブランド、ヴェジャとコラボレーション |

| 2009年 | ・新ライン〈トゥー ビー バイ アニエスベー〉のローンチ |

| 2015年 | ・歌手のジェインが『Come』のミュージッククリップでアニエスベーのシグネチャーである小さな襟付きの黒と白のドレス〈クロディーヌ〉を着用 |

2016年	・アニエスベー創業40周年を記念して、パレ・ド・トーキョーでファッションショーを開催
	・書籍『アニエスベー スティリスト（agnès b. STYLISTE）』をラマルティニエールより出版
	・アニエスが生まれ育った街、ヴェルサイユにブティックをオープン
	・タラ オセアン財団設立。フランスで初めて海に特化した公益財団法人として認定される

| 2018年 | ・デン・ハーグ市立美術館にて開催された偉大なクチュリエを紹介する「ファム・ファタル」展でアニエスベーのルックス3点が展示 |

2019年	・カーディガンプレッション誕生40周年
	・書籍『カーディガンプレッション』をアスリーヌより出版
	・「写真家…アーティストとカーディガンプレッション（DES PHOTOGRAPHES...DES ARTISTES ET LE CARDIGAN PRESSION）」展のカタログを出版（アニエスベー刊）
	・渋谷店オープン

| 2020年 | ・〈ラ・ファブ〉オープン |

| 2023年 | ・日本上陸40周年 |
| | ・京都・祇園にカフェ＆ブティックがオープン |

1941	Agnès Troublé was born in Versailles on November 26th
1958	Married Christian Bourgois Started working at the Jean Fournier Art Gallery on rue du Cherche-Midi in Paris
1962	Agnès designed the agnès b. logo when it was first featured in *Elle* magazine
1966	Designed the famous black and white striped T-shirt featured in the film *Who Are You, Polly Maggoo?*
1973	Registered the agnès b. brand trademark
1975	Opened her first store at 3 rue du Jour in Les Halles district of Paris. The store, located in a former butcher shop, became a pioneering concept store
1978	Opened a new store on rue de Marseille, by renovating a grocery store built in 1910 with wood panelling
1979	Released the snap cardigan
1980	Opened a store in Soho, New York 'Just like with rue du Jour in Paris, I felt like a pioneer.' The collection, embodying agnès b.'s Parisian style, quickly caught the attention of American journalists
1981	Opened the first men's store at 3 rue du Jour Accordingly, the women's and children's store moved to 2 rue du Jour agnès b. launched the kids' line, which was a natural development as she describes working back and forth between the worlds of adults and children
1983	The brand entered the Japanese market and quickly found success
1984	Opened Galerie du jour 'I wanted to create a gallery to show what I love. Though it's called a gallery, it's more about showing the unseen or overlooked aspects of things. Whenever we exhibit paintings, sculptures, and photographs, I've strived to devise new ways to circulate these images'
1985	Awarded the Chevalier de l'Ordre national du Mérite (National Order of Merit) by François Mitterrand
1986	Collaborated with 64 photographers, including Jean-Baptiste Mondino, Gilles Bensimon, Dominique Issermann and Jeanloup Sieff to photograph the snap cardigan The works were exhibited at Galerie du jour in the same year
1987	Opened a store on Fulham Road in London Launched a cosmetics line with L'Oréal's Club des Créateurs de Beauté. In the same year, launched the perfume 'Le B'
1988	Provided costumes for Chet Baker in Bruce Weber's film *Let's Get Lost*
1992	Provided costumes for Quentin Tarantino's film *Reservoir Dogs*. Tarantino, a fan of agnès b., also used her clothing for Uma Thurman and John Travolta in *Pulp Fiction* (1994)
1993	Launched bag and leather goods line 'agnès b. Voyage'
1994	Created the first Artist T-shirt in collaboration with Cuban artist Félix González-Torres
1995	Entered Hong Kong and China markets. Distributed free condoms in stores to support AIDS awareness

Chronicle of agnès b.

1996	Hosted the *Photographers and the Snap Cardigan* exhibition at the Pompidou Centre in Paris
1997	Awarded the Officier de l'Ordre national du Mérite (National Order of Merit) Provided costumes for David Bowie's concert at Madison Square Garden in New York The idea for *Point d'Ironie* was conceived from a conversation with contemporary artist, Christian Boltanski and art curator, Hans Ulrich Obrist
1998	Galerie du jour moved from rue du Jour to 44 rue Quincampoix in Paris
2000	Awarded the Ordre national de la Légion d'honneur (National Order of the Legion of Honour) Launched the 'B Everlast' line in collaboration with the American brand, Everlast
2001	Opened agnès b. galerie boutique in Hong Kong
2003	Purchased a scientific research vessel, naming it Tara, and began the Tara project to better understand the world's oceans and share its results
2006	Celebrated the brand's 30th anniversary with a fashion show and concert at Olympia in Paris Among the guest artists was Patti Smith, who shares a deep friendship with Agnès. To Agnès, a music lover, Patti Smith is 'a legend for young people and a figure with significant impact on fashion' Collaborated with the French sneaker brand Veja, which was only in its second year of operation. Agnès was fascinated by Veja's ethical concept for shoes, its values, design, materials, and commitment to colour
2009	Launched a new line called 'To b. by agnès b.' for young girls and Lolitas, whom Agnès Troublé adores and finds immense joy in designing clothes for: 'I imagine the clothes they would love to wear at that age'
2015	Singer Jain wore the agnès b.'s signature Claudine dress—a black and white dress with a small collar—in her music video for 'Come'. Since then, agnès b. regularly provided her with outfits
2016	Celebrated the 40th anniversary of agnès b. with a fashion show at Palais de Tokyo for which each look evoked the spirit of the brand Published the book *agnès b. STYLISTE* with La Martinière Opened a boutique in her hometown of Versailles Established the Tara Océan Foundation, recognised as the first public interest foundation in France dedicated to the sea
2018	Exhibited three looks at the *Femmes Fatales* exhibition, which introduced renowned couturiers at the Kunstmuseum Den Haag in The Hague, Netherlands
2019	Celebrated the 40th anniversary of the snap cardigan Published the book *The Snap Cardigan* (Assouline Publishing) Published a catalogue for the exhibition, *DES PHOTOGRAPHES... DES ARTISTES ET LE CARDIGAN PRESSION* (Galerie du jour) The photographs in the catalogue were exhibited at rue Dieu Opened Shibuya store in Tokyo
2020	Opened 'La Fab.'
2023	Celebrated the 40th anniversary of agnès b. in Japan Opened a café and boutique in Gion, Kyoto